T0093621

LITHIUM

LUKASZ BEDNARSKI

Lithium

The Global Race for Battery Dominance
and the New Energy Revolution

HURST & COMPANY, LONDON

First published in the United Kingdom in 2021 by
C. Hurst & Co. (Publishers) Ltd.,
83 Torbay Road, London NW6 7DT
© Lukasz Bednarski, 2021
All rights reserved.
Printed in Great Britain by Bell & Bain Ltd, Glasgow

The right of Lukasz Bednarski to be identified as the author
of this publication is asserted by him in accordance with the
Copyright, Designs and Patents Act, 1988.

A Cataloguing-in-Publication data record for this book
is available from the British Library.

ISBN: 9781787385634

This book is printed using paper from registered sustainable
and managed sources.

www.hurstpublishers.com

To my partner, Anna, who always believed in this project, and to my parents and grandparents, who taught me the joy of reading

CONTENTS

INTRODUCTION

If you look at the periodic table, lithium takes the third place out of 118 elements. This hardly seems a compelling reason to read a book about it. But despite its simplicity, with only three protons, this metal is redefining the way we think about energy in the twenty-first century.

Renewable energy produced from solar panels and wind farms has been with us for decades. In 1979, President Jimmy Carter installed solar panels on the White House, to improve its overall energy efficiency, and to promote solar panel use among the public.[1] Until now, the advantage that fossil fuels had over renewables derived from their ability to double as an energy storage medium. By refuelling your tank, you store a lot of energy within a small space that can be harnessed to do work; namely, to power your car. Until fairly recently you were not able to do the same with energy from renewables. Lithium changed that. Lithium-based batteries are the last missing piece in the puzzle of a closed system based on renewable energy.

We are at the very beginning of a groundbreaking change, where energy from renewable sources can be stored to run your car and your portable electronic devices. One day, this green energy will power ocean-going vessels carrying consumer goods you use every day, and let you take off for a holiday without worrying about the plane's carbon footprint. It has not happened yet, but, to paraphrase Amara's law, we tend to overestimate the effect of a technology in the short term and underestimate its impact in the long run.[2]

Around 79 per cent of crude oil currently consumed in the world is used to power cars, planes and ships.[3] The technology required to replace 50 per cent of the oil demand through the use of battery-powered vehicles already exists.[4] The newest electric vehicle (EV) models boast ranges of over 500km on a single charge, and charging times are rapidly improving.[5] Most industry observers have stopped asking whether EVs will replace conventionally powered vehicles. Instead, they are asking when.

The shift from fossil-fuel-propelled engines to the lithium-ion battery is the biggest change in transportation since the end of the nineteenth century, when Carl Benz built the first gasoline engine. It is having a deep impact on industries, national economies, strategic security interests and the struggle to stop or at least minimize the effect of climate change. We are not talking about the distant future, either. The book focuses on exploring the changes that have already taken place as well as those that will happen in front of our eyes, leaving speculation about the world thirty years from now to futurologists.

Large corporations and technocratic authoritarian governments both hold enormous sway in the current political and economic environment, no matter whether we like it or not. The future is being created by government and corporate actors who think in terms of decades—often in the form of five- to ten-year plans, supported by consulting experts' research forecasts. Their targets are being met or not, and progress is more or less carefully evaluated. 'Made in China 2025', a strategic plan worked out by Premier Li Keqiang and his cabinet in the spring of 2015 to transition China from being the factory of the world to a technological powerhouse, turned the EV, battery and lithium industry into a top national priority.[6] In 2016, Volkswagen Group, the world's biggest automaker with over 10.8 million vehicles sold annually, announced its Strategy 2025 plan, aiming to launch thirty models powered purely by lithium-ion battery and expecting 25 per cent of all vehicles sold globally to be purely electric by 2025.[7]

A book like this could not have been written at the time when China and Volkswagen were revealing their strategies to the world. Then, demand for lithium was not high enough to support the element's claim of becoming the 'new oil'. Until 2015, lithium production was more focused on ceramic and glass production than on supplying the battery industry.[8] Nevertheless, brains in the Communist Party of China, or, for that matter, at Volkswagen, must have foreseen the strategic role of lithium when drafting their 2025 plans.

Development of the lithium and battery industry has never lost pace since its early days. The demand for lithium-ion batteries grew by more than 30 times from 2000 to 2015,[9] and is expected to grow by more than another 10 times between 2015 and 2025.[10] Even now, amidst the global COVID-19 pandemic, the companies in this industry make bold decisions on spending tens and hundreds of millions of dollars to build up capacities to satisfy the demand anticipated in five or ten years' time.

They are turning the crisis into an opportunity to gain an edge over more cautious competition, betting on the electric world that is still yet to come.

As the history of the oil industry has been centred on the Western world and the Middle East, with the United States of America playing a leading role, so the lithium industry centres on Asia and Latin America, with a leading role for China. The roles that China, South Korea and Japan play in the lithium-ion-battery industry is yet another sign among many that the centre of economic and political influence is moving from the West to Asia.

The need to secure or maintain sources of hydrocarbons has been driving Western political decisions over many decades, from the encouragement, by the British government, of the adventurist William Knox D'Arcy's search for oil in Persia in the early 1900s to, most recently, the controversies around Gazprom's Nord Stream 2 pipeline. It would be hard to explain the history and current state of the oil industry without references to European, American or Middle Eastern politics. Thus, the book that you hold in your hands will draw substantially on Asian current affairs, to paint the picture of the lithium industry within its Asian context.

As the American version of capitalism—characterized by individual initiative and a free market—left its footprint on the oil industry, so the Asian version of capitalism—collective effort and priorities dictated from the top—is leaving its mark on the battery and lithium industry as it expands.

Even though lithium-ion batteries were first commercialized in Japan, by Sony Corporation,[11] and Japan continues to have a certain edge in the production of key battery components, the book starts in China. China may not match Japan in the quality or elegance of its technological solutions, but it does a great job in bringing the new-energy revolution to the masses, here and now. China is not waiting for battery technology to achieve its peak before starting widespread EV adoption.

The whole system is already in place to serve the industry and customers. Lithium and other key raw materials are mined, processed into chemicals, transformed into components and installed in domestically produced batteries without leaving China's borders at any stage. The batteries go on to power domestic brands of EVs that, as a Westerner, you have probably never heard of. The great thing for the industry observer is that most stages of the production process are carefully monitored by the state, which provides a level of transparency that you would not find even in the European Union's centralized market.

INTRODUCTION

Once an EV is chosen by a customer from the multitude available to suit every pocket, it can easily be charged, and not only in city areas. In 2019 alone, China installed 1,000 new EV charging stations (and that is *stations*, and not just *chargers*) every day.[12] An EV battery, depending on the car's model and usage intensity, lasts between five and eight years.[13] China is booming with new businesses that regenerate a battery without the need to splash out for a new one. Once the battery is really done for, Chinese law states that it must be recycled. China already has recycling capacity to not only deal with its own dead batteries, but also to import and profitably recycle batteries from abroad.

The book tells the story of how China came to this stage of development, both from the perspective of single individuals, who tried to get rich on the wave of economic change, and from that of policy makers, who were striving to deal with high pollution levels and to gain energy security by diversifying the economy away from Middle Eastern and Russian oil—away from places where Chinese political, military and financial clout is still insubstantial.

The opening chapters also introduce the reader to how the lithium and battery business operates on the macro level. For a person born in the twentieth century, oil barrels, tankers, refineries and Saudi Arabian princes exist at the fringes of imagination at the very least. Papers, movies and maybe even some books have surely left you with an idea of the oil industry. It might be wrong or incomplete, depending on your interest level, but it is there. Lithium is a *tabula rasa*. Unless you are especially interested in the subject, you have not dealt with lithium carbonate equivalents and gigawatt hours as units of measure. You do not know what role Chinese lithium converters play or why cathode materials are so crucial for the battery. We will discuss this stuff in a way that is painless even for non-techies before we get to more juicy content.

Chapter Two will deal with China's race to secure the highest-quality lithium resources overseas. China was not strong enough in the past, when hydrocarbon-based resources were up for grabs, either through purely commercial transactions or through political means. But it is not going to miss its chance to occupy the most important seat at the table where the global division of lithium resources is concerned.

The piquant stories of Pinochet's former son-in-law commanding Latin America's biggest lithium producer behind the scenes, as well as of Bolivia's lithium coup, will be unveiled in Chapters Three and Four. They are based on well-documented sources and numerous interviews and conversations I have held with lithium market participants and policy makers.

In Chapter Five, I take a look at lithium and the environment, as well as the net carbon emissions effect of transition to electric vehicles. Lithium mining, as every extractive industry, creates a set of environmental challenges. What sets lithium mining apart from other mining activities is its tremendous water consumption, often in places where water is scarce. In fact, industry insiders call lithium mining from a brine 'water mining'. Chapter Five also touches on the fact that, even if EVs save us from pollutants, especially in inner-city areas where colourless and odourless nitrogen dioxide leaves its grim mark on our lungs, the electric energy to power them, even in advanced economies, comes to a large extent from coal.

Chapter Six is focused on recycling and urban mining. Lithium is a metal that, in theory, could be recycled ad infinitum. It could follow the trajectory of lead, which once came primarily from mines: most of the market demand now is satisfied by recycled lead. The creation of a closed-loop battery economy through so-called urban mining, the harvesting of elements from discarded objects around us, achieved an almost religious status among the masses of mavericks and innovative start-ups, especially in Japan, where it was pioneered at Tohoku University[14] in the 1980s. In 2020, urban mining initiatives have long left the basements and garages of geeks and are pushed forward by governments, municipalities and large corporations. The final chapter looks to the future, discussing projects on the way to make electric flying or ocean-going vessels a reality. It also considers how the transition to EVs affects existing fuel-engine-oriented supply chains. The success of the German economy, the biggest in the European Union, hinges on the automotive sector, and has decades of tradition behind it. Is electrification an opportunity or a threat for brands such as Mercedes-Benz and BMW? The vigorously growing economies of Central and Eastern Europe, such as Poland, Czech Republic and Hungary, profited enormously from supplying the components for the globally competitive German automotive sector. Will they wake up to opportunities created by electrification or will they dwell on their past success? The book tries to answer these among many other questions.

Despite the fact that I am a part of this industry and a cheerleader for it, I try to contain my positive bias. The transition into an electric, battery-driven future is hardly linear. It is not only characterized by continuous advances; on the contrary, there have been many dead ends. These started in the Chinese provinces, so willing to catch up with government targets and so embroiled in internal competition that they forgot about standardization. There was a time when an EV produced in one Chinese province could not be charged in another province because the plugs had different shapes. I could not think of a worse mistake for a product that essentially allows you to move a large distance.

Only a few years ago, European Union decision makers and industry captains were declaring openly that a battery was a commodity. The industry, they disdainfully said, was better left to Asian workshops, where scale and cheap labour provided a competitive advantage. Now Europe is desperately trying to catch up on battery production, funding R&D initiatives and attracting Asian investment in this field within its borders. New lithium mines, developed for a number of years and costing hundreds of millions of dollars, had a history of spectacularly failing a few months short of commercial scale production, due to lithium price volatility and cost overruns. But the dramatic turns in the story of the lithium industry make it all the more interesting to the reader. Moreover, the story continues. Bloomberg, *The Economist*, the *Financial Times* and even the *Daily Mail* write about lithium and batteries on their front pages. I sincerely hope that this book will allow the reader to follow this fascinating industry as it grows in importance with greater understanding and enjoyment.

1

CHINA

A TREND MAKER

China was once a source of innovation: it created paper, gunpowder, the printing press and the compass. When Xi Jinping took power in 2012, he gave a famous speech in Beijing's National Museum of China, outlining the goal of national rejuvenation for China.[1] A key part of this rejuvenation was reverting back to the idea of Chinese people as original thinkers.

China needs innovation to make a qualitative leap and change the structure of its economy. The aim is to transition from being the world's factory to becoming the world's high-tech powerhouse. China's observers who listen to long, ideology-pervaded Communist Party leadership speeches have a habit of counting the words and phrases that are repeated the most, to interpret the message. 'Innovation' happens to be a word that was bounced around by CCP members particularly often in the last years. Another word that statistically ranks high is 'environment' ('shengtai'), used in reference to pollution problems.[2]

The battery is an innovation that helps to solve environmental problems by storing electricity produced from renewable sources and deploying it wherever it is needed. It is also a key element in three out of ten strategic industries that China pushes to develop under the Made in China 2025 plan, which only equals in importance the more publicized Belt and Road Initiative. These three strategic industries are green energy (including green vehicles), power equipment and new materials.[3] Two of the three are self-explanatory, while new materials may require some additional clarification. The modern lithium-ion battery would not be possible without a new generation of chemicals. Cathode and anode materials constitute the heart and soul of the battery, guaranteeing the level of performance that was beyond reach of EV manufacturers only a decade ago. The Chevrolet Volt, a brainchild of General Motors, was supposed to start the EV revolution in the US as early as 2011.[4] If the Volt had been successful, this book would not start with China. Mass sales of the Volt would have spurred the development of battery materials in the US. The lithium mines that were supposed to be developed by 2020 in Nevada, North and South Carolina, South Dakota and California would already be there, running at capacity and spitting out the new oil. It wouldn't be only Californians' privilege to enjoy a dense network of chargers.

But what the Volt lacked was an advanced battery chemistry, the cathode and anode material that would allow it to travel more than 40 miles (64km) on a single charge.[5] According to the AAA Foundation for Traffic Safety, Americans drive on average 31.5 miles a day.[6] The 40 miles limit for the Volt triggered a range anxiety that still lingers in the psyche of US consumers, even if current driving ranges of up to 500km (300 miles) seem to be sufficient for EV mass adoption.

In China, the electric revolution took place gradually, rooted not in widespread adoption of EVs but in the increasing popularity of electric bikes.[7] Katie Melua's inspiration for the song 'Nine Million Bicycles in Beijing', which became a hit throughout the world in 2015, came from her visit to the city.[8] For a long time, visits to China, much like visits to Vietnam nowadays, left an impression of cities humming with bikes, the affordability of which made them more popular than cars.

The beginnings of the electric bike industry can be traced back to the 1960s, when it was supported by Mao. Surprisingly, it found its niche among other segments of the centrally planned economy, which was otherwise focused on heavy industries such as coal, cement, fertilizer and steel production.

Nevertheless, electric bikes in the 60s were not very successful. The end of the 70s and the beginning of the 80s was a time when the imbalance between the level of development of the light and the heavy industries in China started to be acutely felt. Beijing was often flooded with acid rain, thanks to the concentration of cement and steel production in its vicinity, while within the city itself, there wasn't as much as a pencil factory.

For the first time, the party allowed enterprises to address market needs outside of centralized plans. In this ambience of opening, dreams of Chinese-produced electric bikes were revived, but fell flat after some initial engineering efforts due to lack of components. The project also received a lukewarm reception from the government.

Early electric bikes were not powered by lithium-ion batteries, as these were only brought on the market in the 90s, by Sony and Asahi Kasei, a chemical company. Instead, they ran on lead acid batteries, the same that help to ignite engines of conventionally powered cars. But the idea of a battery-powered means of transportation was planted in the Chinese engineering community, and was clearly supported by underlying strong consumer interest. It waited for more economic freedom and better battery technology to bloom. The leap in performance from lead acid to lithium-ion batteries was gigantic. The range of 32 km (20 miles) could be achieved on a battery that was six times smaller. The growth in sales volume reflected the technological improvement, with sales skyrocketing from 56,000 units in 1998 to over 21 million in 2008.[9] The demand generated for lithium-ion batteries through successful bike sales created revenue streams for companies, which amassed capital to later turn into EV and battery powerhouses.

The car market has also been growing explosively in China, ever since the mid-80s. Even if China's GDP per capita in 1985 was a meagre $294, car imports have been booming, especially those from Japan.[10] Much like other communist countries, China, despite the poverty of most of its citizens, still possessed a wealthy elite mostly connected to the Party. After China spent $3 billion on cars in 1985, its leaders started to worry about a deficit induced by car imports.[11] They introduced a set of measures, starting from stricter currency controls to make it more difficult to buy cars in yens or dollars, and ending with a decree almost completely prohibiting car imports for a period of two years.[12]

While deploying defensive measures, they counter-attacked to benefit China's own automotive industry. But since the technology was not there to even build decent electric bikes, they started a process that continues to this day in some segments of high-tech, where Chinese capabilities are not yet on par with the rest of the world. Foreign automakers, seeing the success of Japanese in China, were salivating at the thought of accessing a market of 1 billion people. Heavy import restrictions checked this dream. But there was a way out: if you wanted to sell cars in China, you had to produce them there, in joint venture with Chinese partners. Companies such as Volkswagen, Citroen, Peugeot and DaimlerChrysler went for the opportunity. This type of joint venture started the know-how transfer to China, even if Western automakers protected themselves as much as they could to keep Chinese partners in the dark, for instance through assuming a strategy of importing ready-made parts kits which required only a simple assembly on site.

The strategy, however imperfect, worked. It has since been successfully repeated many times over, for instance in the solar and wind energy generation industries. But the automotive industry has been a pioneer in this scheme. China still tries to apply it now, with semiconductors or with AI, but the world has become more cautious of the trade-offs involved. The US started to refer to this practice during the Trump administration as stealing trade secrets, and it became one of the leading causes for a trade war. China continues to see it as a transparent and fair arrangement: it is the price of accessing China's enormous market, with its growing demand. But the recipe for China's success in spurring their own industries does not end with sharing arrangements. The other part of the equation is state aid. Once the technology transfer is complete, China starts its systematic financial support to the whole industry, support from which even China-based foreign firms tend to be excluded. Considering how deep Chinese state coffers are, the strategy allows China, in time, not only to drive out foreign competition from the domestic market, but also to corner them on the global stage. This is precisely what happened with solar panels. At some point, China started to require that its big municipal solar projects only use solar panels that are at least 80 per cent made in China.[13] Foreign companies moved to produce in China, establishing joint ventures and sharing the technology. Then China started to heavily subsidize its own solar companies, including production for exports. Now, eight out of the world's ten largest solar panel manufacturers are Chinese, and China holds over 60 per cent[14] of the global market share.[15]

Most of the automotive companies that achieved early success in China were not privately held. Chang'an Automobile Group, now belonging to so-called 'Big Four' domestic automakers, is a spin-off from a defence company[16] with long traditions, having supplied arms during the Sino-Japanese War in 1937. Chang'an continues to operate as a wholly owned subsidiary of China Weaponry Equipment. After achieving success with conventionally powered cars, it pledged to completely phase out their sales in 2025, and to turn 100 per cent electric.[17] State-owned arms companies selling eco-friendly EVs might baffle Western Sustainable Funds managers, but this does not raise any eyebrows in China. Another important player on the Chinese automobile market, Jiangxi Changhe Automobile was until 2010 a subsidiary of Aviation Industry Corporation of China, maker of military aircraft.[18] Changhe learned to make cars through a joint venture with Suzuki. Even now, if you look at Changhe's newest models, some see 'Suzuki' influences. Changhe is also turning electric, but with less zest than Chang'an, as it is still hoping to conquer frontier markets such as Myanmar, Laos and Nicaragua with its gasoil and diesel-powered offering. Another large automaker, Hafei, also started out as a subsidiary of the same military aircraft company, so there is a pattern here.[19]

The rest of China's automakers that accomplished an early success were or still are state-owned enterprises, with the notable exceptions of Geely (which started as a fridge maker)[20] and Great Wall Motors. Chery, with close to half a million sales in 2019, was co-founded by a group of apparatchiks from Anhui province in the late 90s. Until 2003, it had in fact been operating illegally, as it did not have the required license to produce passenger cars.[21] The *Wall Street Journal* in 2007 described its organizational culture as 'an odd hybrid of Communist state enterprise and entrepreneurial start-up'.[22] Chery has received a lot of criticism for allegedly copying elements from other auto-making brands. GM-Daewoo even took Chery to court, which resulted in a high-profile case where the vice-director of China's State Intellectual Property Office publicly defended the company.[23]

It is not surprising that in a country where opportunities to get investors had been strictly limited in the past (as there were no venture capital funds in China in the 80s or early 90s), capital intensive projects, such as building cars, were likely to happen only in strict cooperation with the government. The defence sector has been a driver of innovation in the United States and pretty much everywhere else, and building cars at that time in China was in fact innovative.

The development path of the Chinese automotive industry is typical for capitalism with Chinese characteristics. First, the party sees a strategic need to develop a sector of industry to address macroeconomic imbalances. If knowledge transfer is needed, a mix of coercive and incentivizing legislation is drafted to help execute it. Once the know-how is there, subsidies kick-in. But since so many of the key market players are state owned or affiliated with the party either through politically involved executives or through financing dependent on the state, the decision to involve the company in a state-promoted industry is not simply based on economic calculation.

Anywhere else in the world, large corporations evaluate new projects based on opportunity costs and benefits as well as on internal rates of return. But the 'Chinese dream', unlike the American dream, is, as Xi Jinping pointed out, collective, and the executives of state-owned companies need to factor this into their investment decisions. It allows China to move fast and change the structure of its economy according to the government's vision. Naturally, such a top-down approach to economic development also results in excess capacities, market bubbles and sometimes products of questionable quality. China's EV, battery and lithium industries have also not managed to avoid these types of problems.

The government, seeing the success of electric bikes on the market, was even more eager to bet on the development of the EV industry early on. There were a number of good reasons to do that. Besides reducing pollution in the cities, the development of the EV industry in China had the potential to spin-off a whole new segment of the economy in which China could get a competitive advantage on a global scale. This would start with mining and chemical processing, but also involve pushing new technologies forward, such as lithium-ion batteries or autonomous driving. The development of a new energy industry can be traced back to the 863 programme. On 3 March 1986—86/3 in Chinese notation—a group of four Chinese physicists sent a letter to Deng Xiaoping. The four were Wang Daheng, Wang Ganchang, Yang Jiachi, and Chen Fangyun,[24] who had made their names in dual (civilian plus military) use research, including nuclear and satellite applications. Deng Xiaoping was China's true statesman, who is often referred to as the architect of modern China. He is perhaps best known for China's Special Economic Zones experiment, in which he introduced a market economy and foreign investments in specially designated coastal areas—thus in effect sandboxing a capitalist experiment in China. Its success, which included cities such as Shenzhen, paved the way for countrywide economic reforms.

Within the letter, the scientists outlined a 'State High-Tech Development Plan' to stimulate the development of advanced technologies in several focus areas in order to make China independent from economic reliance on foreign powers. Deng Xiaoping was so enthusiastic about the plan that he made a decision to endorse it within two days. He wrote to his party colleagues: 'Quick decision should be made on this matter without any delay'.[25] History books often state that the plan had been inspired by Ronald Regan's Strategic Defense Initiative proposed in 1983, later dubbed the 'Star Wars' plan. But this is a very Western-centric view, as any direct comparison between 863 and Star Wars seems far-fetched. Star Wars was about the development of an anti-missile system to protect the US from a nuclear attack, especially from the Soviet Union. It gave stimuli and funds for development of many cutting-edge and often fantastical technologies, such as laser beam weapons. But despite its conducive effect on the high-tech industry in the US, it had one well-defined goal and a specific military application.

This is very unlike the 863 plan, which starts by acknowledging that China is a developing nation and thus cannot allow itself to disperse its scientific focus. It further declares that there are only a number of focal areas where China should spend the funds and develop talent to close the gap with the outside world. 863 goes on to state that China stands a chance to position itself as a leader in entirely new areas of technology, where competition is not yet so strong, if it enters them soon enough. Among seven initial areas, two are directly related to battery production—new materials and energy.[26] The 863 plan was scheduled to run in nine five-year periods, for a total duration of forty-five years. It ended in 2016—earlier than intended—while being replaced with a number of other initiatives. The 863 plan concentrated on advancing basic research, with potential for commercial applications. This is rather interesting, as it contradicts the early image of China as a country that just wants to copy existing solutions to monetize them. Japan's high-tech industry still stands strong due to its investment in basic research, which led to the creation of many patents in the niches that serve as building blocks in battery, display or semiconductor industries. It seems that China had a similar ambition for chosen areas of its economy to rely strongly on basic research, and that it wanted to go about it in a methodological, state-supported way. Thus 863 was a purely Chinese invention.

In the first year of its existence, it received 10 billion RMB, equivalent to 5 per cent of all government spending for that year.[27] EVs were included as a focus area in 2001, and basic research started on development of pure and hybrid vehicles, especially targeting problems related to power batteries and electric motors.[28] Once the groundwork in basic and applied research had been firmly established, China followed with ordering sixteen state-owned companies to form an electric vehicle industry association in Beijing, as a platform to exchange know-how and foster the industry's growth. Member-state-owned companies pledged to invest $14.7 billion in development of the EV industry. This decision was not motivated by a market need or even by those companies' own vision of the future. Instead, the companies put that much money on the table so as to comply with a vision of the party's higher echelons for a greener, less oil-dependent China. The government's plans have been very ambitious from the very beginning. In 2008, China already had a goal for 10 per cent of its fleet to run on alternative sources of energy by 2012. The Olympic Games in Beijing in 2008 presented an excellent opportunity for China to present its progress on EVs. Nevertheless, in 2008 EV sales numbers were still well below 1,000 units, and during the games, the Chinese did not manage to brand themselves as a leading EV nation of the future.[29] Remarkably, the rise of China to the rank of the world's biggest EV market went unnoticed in the West for a very long time. In 2008, the Toyota Prius and the first Tesla models got all the attention. If anybody were to bet on the largest EV market in the next decade, the odds were strongly in favour of the US or Japan.

The biggest boost for the Chinese industry, though, came only a year after the Olympics, under the 'Ten Cities, Thousand Vehicles' programme.[30] This was despite the fact that the assumptions for the plan were highly questionable. Instead of supporting the top three or top five companies developing EVs for the whole country, the government assumed a decentralized approach. Funds were to be distributed to selected cities and a goal was set to get 1,000 vehicles on the road in each of them. The details on how this was going to happen were left to municipal powers. The plan ominously resembled Mao's Great

Leap Forward: Mao's 1958 plan to move China from an agrarian to an industrial economy. He personally saw the level of each country's industrialization as strictly correlated with the level of steel production. This meant that he was hell-bent on catching up in steel production numbers with the world's most industrialized nations at the time. Instead of supporting the largest steel mills in the race, he decentralized steel production to an extreme, encouraging average citizens to produce steel in backyard furnaces. Obviously, the plan backfired, as the quality of produced steel was good for nothing.

But the tradition of zoning social and economic experiments with a potential of introducing major changes into the society also has more positive connotations in China. Deng Xiaoping's Special Economic Zones on the coast, for example, are a reminder that this approach has been around for a while, and has had some successes. The geographically limited pilots allowed for new solutions to be introduced and safely tested and evaluated in what amounted to a macro-laboratory, before they would be rolled out nationwide.

The selection of cities for pilot programmes in China has been far from random. Pilots have typically been introduced in cities or provinces known for a low level of political resistance, with economic characteristics suitable for the pilot's objectives, and that were representative of the country at large. More risky pilots have also traditionally been introduced in places far away from Beijing, so that the risks of political backlash in case of pilots going wrong were minimized.

Zoned pilot programmes are not a thing of the past, either. At the time of writing, and the Covid situation permitting, pilot zones for cross-border e-commerce are being set up.[31] More and more foreign residents are buying directly from Chinese e-commerce platforms, as this helps to secure good bargains. E-commerce companies from pilot zones will receive tax benefits and will be supposed to work together, for instance by sharing warehouse space outside of China. China's infamous Social Credit System, where people are scored for 'good behaviour' such as donating blood, as well as 'bad behaviour', such as making reservations at restaurants and not showing up, was initiated as early as 2009 in regional pilot zones. Usage of pilot zones is highly reminiscent of Deng Xiaoping's motto for Chinese development—'crossing the river by feeling for the stones', i.e. introducing changes cautiously and gradually. The success of pilots is evaluated on a continuous basis, and changes to initial goals and methods are made on the go, according to the feedback received. Any successes of pilots are highlighted in the mass media. The aim is to make people champions of the changes, so that the gradual nationwide roll-out of new ideas would ideally happen not through top-down coercion but thanks to grassroot support. Since the political risks were low in the case of EV implementation, and since the pollution levels were high in China's flagship cities, the '10,000 vehicles for 10 cities' pilot included Beijing and Shanghai from the very beginning. Soon after the pilot started, each of the ten cities figured out its own unique approach to the programme, by playing on its existing strengths. Capitalistic Shanghai relied on its high level of private investment, bureaucratic Beijing on creating incentives through taxes and regulation, and innovative Shenzhen on cooperating with its strong high-tech companies such as BYD. Chongqing, home to the world's largest hydropower project, the 'Three Gorges Dam', started working on fast-charging batteries and fast chargers, to leverage its cheap renewable energy and robust electric grid.[32] So through the feedback loop and that pragmatic approach, the goal changed from a purely quantitative (10,000 vehicles) to a more qualitative orientation, where each city contributed to different aspects of the national new energy revolution.

By 2012, three industrial campuses had been established in Beijing, where companies had been working hand in hand with scientists to move battery and EV engineering forward. Thanks to the private-public partnership, around 200 fully electric taxis were in operation. Taxes were made substantially lower for EVs, and, most importantly, it became easier to secure a license plate for an electric car. In China's biggest cities, money is not the main problem to keep you away from your dream car. Since 2011, due to congestion and pollution levels, China operates a system of annual quotas for new license plates. Since there are more applicants than new license permits to be awarded, a bimonthly lottery determines who will be able to register a new car. Most people wait for years to be able to acquire a vehicle. In 2020, the quota for passenger vehicles in Beijing was at 100,000 units—40,000 for conventionally powered cars and 60,000 for new energy vehicles.[33] The ratio between conventional and electric cars in new license plates quotas nowadays shows the long way that Beijing has travelled in the last eleven years in regard to electrification of its car fleet. In the beginning, it was much easier to get a licence for an electric car than a gasoline fuelled one. Nowadays it is still easier, but some who are less lucky still have to wait for years.

At the beginning of the EV industry, battery costs were much higher than they are now. In 2010, 1 kilowatt hour (kWh) of battery pack cost $1,100.[34] Nowadays, the cost is around $150 to $160 for 1 kWh. The simplest way to imagine that amount of energy stored in a battery is to think about a single 100 watt lightbulb. If you keep it switched on for 10 hours you are racking up 1 kWh of energy. The cost of the battery pack was the most prohibitive factor in a massive EV adoption. So the pilot cities addressed this issue by supporting schemes where the battery pack was leased to the customer instead of being sold.

In Shanghai, Beijing, Shenzhen and Chongqing the pilot city programme was quite successful, even if targets as expressed in absolute numbers were far from being met. By the end of 2012,

Shenzhen's target was to have 4,000 vehicles on the roads, while little over half of that number was achieved. Still, in terms of sheer number of units, Shenzhen was the most successful. Shanghai assumed a more modest goal below 2,000 units, which was met at around 70 per cent. Some cities, such as coastal Tangshan in the vicinity of Beijing, were a spectacular failure—with a target near 2,000, below 100 vehicles were on the road in 2012.[35]

The programme was a success in the sense that it popularized EVs early on. Pilot cities managed to convince their inhabitants and later on the rest of China that the electric future had started. At the same time in the US and Europe, electric vehicles remained a choice for well-off geeks and upper-middle-class Californians.

That is not to say that the Chinese pilot was not fraught with pathologies. Local officials often touted their success, so as to keep looking good in front of comrades and attracting further capital inflow, with very little substance to show for it. In 2012, there were already twenty-five cities in the programme. The participation of some of them was questionable from the very beginning. Hohhot in Inner Mongolia, for example, despite the proximity of China's largest rare earths mine (producing elements used in EVs' electric motor), had a very small industrial base and not much of a socio-economic environment to work with. Perhaps the most telling fact is that Hohhot's first EV charging station opened only in February 2018.[36] At some point, the initial ten cities became so fed up with the trumped-up rhetoric of some of the new participants in the pilot that they initiated a committee to check on each others' progress. Local authorities also started to dabble in regional protectionism, to increase their statistics. One of the strategies was to subsidize sales of locally produced EVs while not subsidizing models produced in other cities. This, made otherwise very competitive BYD models, produced in Shenzhen, for instance, impossible to sell in Beijing. Naturally, that localized approach was harmful for the development of real national champions. Also, standards were introduced in isolation and compatibility was limited on purpose. For instance, you could not charge an EV from one megacity in the other megacity, as the sockets were in different shapes.

The mastery of the battery was key to producing competitive cars, since this is what determines an EV's driving range, speed of charging, acceleration and safety. But as the story of electric vehicles in China started with bikes, so the story of batteries started with mobile phones. It was their affordability that allowed mobile phones to spread across China, before laptops and other portable electronic devices. But in early 2000, when the mobile phones manufacturing boom reached its peak, China was still strongly relying on imports of Japanese batteries to power its handheld devices. Japan was first to commercialize lithium-ion batteries and, due to its strong basic research and use of robots in their manufacturing, kept the lead in the battery industry worldwide. Thanks to this, Japan was able to capitalize strongly on the Chinese electronics manufacturing boom. China's rechargeable electronic gadgets simply would not work without Japanese batteries in them. The main problem for Chinese companies in entering the battery industry was the know-how and necessary capital expense just to get started. Basic, automatized Japanese production line cost started from $100 million.

It was a visionary from Anhui province who saw the other way forward. Wang Chuanfu, who had a background in chemistry and material science, decided to enter the battery industry against the odds, naming his company BYD—Build Your Dreams. He reportedly started by buying Japanese batteries and reverse-engineering them with his colleagues in academia.[37] He also looked at Japanese patents to better understand the technology. In China at that time, you could buy pirated movies, music and books in legitimate shops on the high streets. Intellectual property laws were the last thing to worry about. Once Wang and his team figured out how commercial-grade batteries operate, he decided to build his own production line. He replaced expensive Japanese robots with people: both skilled and unskilled labour was cheap in China when Wang was starting out. And even in Japan, some tasks at the production line had to be completed manually.[38]

The work must have been boring and simply unhealthy, and the turnover rate at the company was high. But Wang's batteries sold well, retailing at $3 per cell while Japanese batteries were trading at $8.[39] Human labour turned out to be significantly cheaper than automated production lines—in terms of both capital and operating expenses. The beginnings of BYD might make grim reading to today's readers, being reminiscent of deplorable labour practices straight out of England's industrial age. But BYD's founder is an icon in China now, who, with his bootstrapping can-do attitude, became one of the fathers of the new-energy revolution. BYD, for its part, is currently a multibillion-dollar company, with Warren Buffet as one of the major shareholders.

The success of BYD was not only based on reverse-engineering Japanese cells and on cheap labour, but also on its continuous ability to transform and to keep riding a wave of strong demand. From its humble beginnings as a cottage producer of batteries for cell phones, the company made it to the top three biggest automakers in China. Its success did not go unnoticed in the West. In 2010, *Bloomberg Business Week* ranked BYD as the eighth most innovative company in the world, together with decades-old giants such as Ford and Volkswagen, who spent billions of dollars on research and development.[40]

The ability to move to adjacent or completely new segments of the market, where the demand is growing, is a unique characteristic of Chinese entrepreneurship. Western business schools put emphasis on specialization and on investing in core capabilities, both for small companies and at the corporate level. The Chinese are more pragmatic, and move quickly where the market is, even if it means that they would need to learn the ropes from the very beginning, and that the quality of the initial products would be far from perfect. Perhaps there is no better example of this mentality than Ningbo Shanshan, which now is one of the top battery materials producers in China. In 2006, 93 per cent of its revenue came from apparel.[41] Men's clothing, especially business suits, was the segment where Ningbo Shanshan made its first major profits and built capital. Fast forward ten years, and battery materials account for 75 per cent of Ningbo Shanshan's revenues.[42]

BYD, perhaps, did not make as great a leap into uncharted waters as Ningbo Shanshan. But diversification was still the name of the game. BYD has grown into new segments vertically. It started from selling batteries to foreign mobile phone makers who were using different Chinese suppliers for different parts of their mobile phones. The way it worked was that each Chinese factory received specifications for different electronic components, which would then be assembled by yet another party. The problem was that even the tightest specification in the electronic industry must have some degree of tolerance included for component characteristics. In most cases, this works just fine, but if there are too many different suppliers, and each of their products deviates from the specs, even if within contractually acceptable brackets, the end product may not work as well as expected. Added to this, a supply chain that is overly fragmented increases the risk of delays. BYD saw this problem, and thus proposed its cell phone battery purchasers to contract them for more than just batteries.[43] This is how BYD became a cell phone manufacturer.

Within ten years of its founding, BYD already had more than half of the world's battery market. The company became the world's fourth largest battery manufacturer and the largest player in China.[44] Seeing the growing EV market and the government's support for the e-mobility sector, Wang Chuanfu realized that he should not stop at expanding his cell phone operations. Indeed, the battery, where the company had the strongest expertise, was a building block determining EVs' key performance parameters. In 2003, BYD acquired the state-owned Tsinchuan Automobile Company. The company knew how to make batteries, but had to learn very fast how to build a car, and the shortest path to do that was through acquisition.

By 2013, not less than 51 per cent of BYD's revenues came from automobile sales. The company's F3 model became a bestseller in China even earlier, just seven years after BYD got into the automotive industry. The first successful BYD models were not even electric. So success, at first, did not come so much from mastering the battery, but from understanding consumer tastes. In interviews Wang Chuanfu often referred to the concept of 'mianzi'—face.[45] He understood that in China cars not only have a utilitarian value, but are first and foremost a conduit for prestige. Even the design of an affordable car needs to reflect that. He also believed that in China you typically carry more people in a car than in the West. Thus, the backseat needed to be very spacious and comfortable. The bestselling F3 model reflected these characteristics. The BYD founder, despite occupying China's rich list by now, still understood customer's wallets. Chinese middle-class income is still much lower on average than that in Western Europe, the US or Japan. BYD has been offering cars from different segments, from compact ones to large sedans, at price tags between $4,400 (!) and $15,000.

Despite a tremendous edge in battery technology, BYD's first electric car was not much of a success. BYD's F3DM ('DM' standing for dual mode), which provided a possibility to switch from gasoline to electric propulsion on the road, left the production line in 2008. But in its first year, BYD sold only forty-eight units, and not even on the free market but to government agencies and state-owned companies, which were under pressure to support the project.[46] Production of the F3DM ran until 2013, and its total sales closed below 3,500 units.[47] There were two roadblocks to the success of Chinese early electric cars. The charging network was still underdeveloped, and in urban areas, people rarely lived in one-family houses, where you could easily leave the car plugged in overnight.

The second obstacle was the price. The purely gasoline version of the F3 was actually a success—possessing the exact same design and features it cost the equivalent of $8,750, while an electric version only started at $21,900. For the Chinese scrambling middle classes, it was a no go. In 2010, the government rolled out subsidies for electric vehicle makers in cities that were part of the pilot programme discussed earlier. Shenzhen-based BYD was covered under the programme and received around $7,600 from the government for each model it sold.[48] The prices in dealerships were lowered accordingly. Later on, additional subsidies were introduced on the consumers' side, for private purchases. But the government support came too late to render the F3DM a success.

The F3DM had a relatively weak battery in comparison to what is state of the art now. The battery pack capacity was at 16 kWh, more than four times smaller than the newest Tesla model, and this allowed for no more than 60 km (37 miles) of driving range in the electric mode.[49] A tank was able to add another 480 km (300 miles) on gasoline. The car used a Lithium Iron Phosphate battery—usually just called an LFP (Lithium Ferro-Phosphate) battery. The chemistry in most batteries, such as LFP, has been refined over the years, meaning they are still used now, in the newest generations of electric cars. To simplify, the chemistries most commonly used in lithium-ion batteries are LFP, NMC (Nickel-Manganese Cobalt), NCA (Nickel Cobalt Aluminium) and LCO (Lithium Cobalt Oxide). NMC and LFP still fight for market share among power battery (EV and Utility Energy

Storage) applications. They have their own advantages and disadvantages that we will be discussing throughout this book. For the last two to three years it seemed obvious that NMC batteries would be the winner, at least for EV applications, due to their generally superior performance parameters. But the refinement of the LFP battery continued, to a vast extent due to BYD pushing the frontier of research. BYD has no sentimental attachment to LFP: it is the cheaper and safer option. It does not contain cobalt, which is expensive and subject to volatility in price. Cost advantages speak to more price-conscious consumers in emerging markets, while the absence of cobalt might convince clientele aware of the negative impacts of cobalt mining. BYD found their first success on the electric market with the Qin model, a successor to F3DM, named after the first dynasty of imperial China. In the spring of 2016, its sales passed the 50,000 unit mark.[50] For two years it was also the best-selling electric model in China. Its battery was the same LFP type, but a newer iteration, with higher energy density. The model that sold so well was still a hybrid, able to go only 70 km (43 miles) on battery alone. The purely electric Qin arrived only in March 2016, three years after the hybrid premiere. Under the series Qin EV300, it boasted a 300 km (186 miles) driving range. Without the subsidies, prices started at $36,600. With subsidies applied, the price has been reduced by a staggering $17,000 dollars, almost half of the car's value.[51] The example of the Qin EV300 is a very good demonstration of the amount of financial support that the Chinese government is willing to offer to get this industry off the ground. The biggest stimulus package ever introduced outside of China is a German scheme announced only in 2020, also as a part of post-Covid stimulus efforts. Its maximum of EUR 9,000 ($10,150) still pales in comparison to Chinese levels of support.

BYD's method was not the only way to make it in the Chinese battery business. Tianjin Lishen Battery took a completely different approach to succeed. While BYD's beginnings were based on tinkering and bootstrapping, Tianjin Lishen had the full might of the state behind it, as well as a starting capital of $272 million. As a state-owned player it did what many private battery entrepreneurs could not dream of: it bought a cutting-edge battery assembly line from Japan.[52] While BYD's workers were toiling away at the production line, Tianjin Lishen had a fully automatic assembly line, and, just a few years later, partnerships with Motorola and Philips. While the amount of waste at BYD was reaching 30 per cent of total output,[53] Tianjin Lishen had been very efficient and produced high-quality cells.

Battery makers' capacities are measured in gigawatt hours (GWh). Betting on quality and technology does not always make you win on the open market, despite what business school textbooks say. In 2019, BYD had 28 GWh manufacturing capacity and, together with CATL,[54] led the power battery market in China.[55] Lishen, with 10 GWh, still counted as one of the major players, but was far from the podium.[56]

To power the battery boom China needed lithium. Lithium is the only element in lithium-ion batteries that exists in all battery chemistries, be it LCO, NMC, NCA or LFP. Other elements come and go. One can find nickel in NCA and NMC, but not in LCO and LFP. The same goes for cobalt, which can be found in LCO, NMC and NCA, but not in LFP. Lithium is everywhere and consistently constitutes around 10 per cent of the cathode material across all the battery chemistries.[57]

In the early stages of the development of its battery industry, China relied on the country's vast interior for its lithium needs. Soon enough this proved not to be enough to power the battery boom, and China started to look to secure lithium assets abroad. The story of domestic exploration is not less interesting, and it teaches us something about China's history that is relevant to the country's current affairs, and provides us with deeper insights into China's attitudes toward Xinjiang region, infamous for a level of state surveillance and Uyghur re-education centres.

If you look at the map, Xinjiang is situated in the north-western part of China. The province borders the 'stans' (Kazakhstan, Tajikistan, Kyrgyzstan) and Tibet. In the past, Xinjiang had a long border with the Soviet Union, as the 'stans' were a part of it. The province is full of open spaces, deserts and grasslands, enclosed by mountain ranges. The landscape is breathtaking but also hostile. It is perhaps best summed up in the words of Xinjiang party secretary Wang Enmao, in 1954: 'there is very little of anything above ground in Xinjiang, many areas are simply barren land—but buried below is a limitless supply of treasures'.[58] Indeed, Xinjiang has ample reserves of oil, as well as non-ferrous, minor and precious metals, with lithium among them. In Mandarin the name of the province means 'a new frontier', an apposite term, considering that Xinjiang has been annexed to China by the Qing dynasty only at the end of the nineteenth century. At first nobody realized the value of the treasures underground. Xinjiang's importance was mainly seen in terms of geopolitics, as a sort of nexus between the great powers active in Eurasia—China, Persia, Russia and Europe. Its location meant that it remained very difficult to control by Beijing. It was susceptible to foreign, especially Russian influence. The period of turmoil in China, starting with the deposition of the last Qing emperor, through the civil war, Sino-Japanese war and finally the establishment of the Communist People's Republic of China in 1949, did not help with Xinjiang's integration. The province's governors had been performing a difficult balancing act between the Soviet Union, the Chinese Communist Party, the Kuomintang and political ambitions of the local Muslim populace, who at some point managed to establish a short-lived East Turkestan Republic. The province's assertiveness was not only fuelled by its leaders' ambitions. Despite its size, the province was very poor, while economic transfers from Beijing were patchy, and at extended periods of time non-existent. Forty thousand people living in Yan'an, Shaanxi province, one of the poorest places in China, were still producing more in tax revenue than the four million people of Xinjiang province.

To keep its populace from starving, local powers were ready to strike a partnership with anybody who was willing to buy the province's output. At first this was furs and wool, and later on natural resources. Due to geographical proximity and better trading routes, the Soviet Union became the republic's natural main trading partner.

Beijing was worried about Soviet influence, to the extent that in 1938 Mao Zedong's brother, Mao Zemin, was appointed as Xinjiang's treasurer. In a letter to Mao Zedong, he wrote, referring to province governor Sheng Shicai's propensity to take Soviet loans, 'We still do not even know how many other outrages to both heaven and earth he has committed'.[59] The Soviets provided what Beijing was unable to give—capital, engineering expertise to develop mineral resources and access to global commodity markets. Sheng Shicai was taking monetary loans from Moscow, and these were returned in kind. He even paid visits to Stalin himself. In 1938, during one of those meetings, Stalin asked about Xinjiang's tin resources. Sheng replied that there was tin in the province, but that he did not have a plan to extract it.[60] It was enough for Sheng to mention the presence of tin to receive the Soviets' full support in its extraction. With that kind of relationship and a growing industrial Soviet base, hungry for mineral resources to fuel its growth, especially in the face of military efforts in the Second World War, an extraordinary close cooperation quickly followed. Xinjiang provided Stalin with oil, gold and rare metals with important military applications, such as beryllium, tantalum, molybdenum and tungsten. The province was ideally placed to support the Soviet war machine, as it was far from enemy lines—especially German and Japanese. Seeing this advantage in the 40s, the Soviets were willing to pour even more money and talent into the province, hammering out the infrastructure that the Chinese Communist government would build upon after the war. The plan was not without its risks. Even today, with all the technology we have at our disposal, mining is an extremely risky business. Analysts estimate that only 2 per cent of all junior companies exploring for gold end up building a mine. In fact, the geologists and entrepreneurs of the Qing dynasty were also excited about the prospect of mining for gold in Xinjiang's Altai mountains. They quickly became discouraged by the results of the exploration effort as the short-lived mining investment bubble burst.

When needed, the Soviets were also willing to back up Sheng with crude force, to protect their sphere of influence and resource base. As are many places in the world whose borders have been artificially carved out, Xinjiang was ripe for ethnic conflict. In fact, Xinjiang contains two historically and ethnically disparate realms, separated by the Tianshan mountain range. Dzungaria, to the north, is inhabited by the Buddhist Dzungar people, while the Tarim basin is inhabited by the Muslim Turkic people who speak Uyghur. The aspirations for independence of Uyghurs were thwarted with the use of the Soviet military, often operating undercover, a tactic reminiscent of today's Russian covert operations in Syria, Libya or Ukraine.

Sheng Shicai's Soviet preferences were not inspired by ideological sympathies but by realpolitik. From 1942, when Stalin appeared to be losing to the Nazis on the Western front, and when NKVD purges of Soviet allies made Sheng increasingly worried for his personal safety, he started shifting alliances. He teamed up with the anti-communist Chiang Kaishek, who was most eager to make use of Xinjiang's oil to propel the war effort, and of tungsten, which Americans had already been importing from China—all while paying in hard dollars.[61] Chiang Kaishek is a towering figure in Chinese history, a leader of the Nationalist Party after Sun Yat-sen as well as a talented military strategist who briefly managed to unify China before losing in the civil war to Mao Zedong's communists. The task for Sheng Shicai was difficult: as a result of his long pro-Soviet politics, he feared for his position and his life. What kept him afloat was his ability to, on the one hand, replace pictures of Stalin with pictures of Chiang Kaishek in public buildings, while at the same time keeping Soviet expertise and investment in the natural resource sector flowing.[62] This was necessary, as the Kuomintang had neither money nor expertise to mine and market their mineral resources fully independently. However, the balancing act did not last long. Chiang Kaishek's enthusiastic politics of opening the new frontier and reconnecting Xinjiang to the mainland turned social attitudes against the Russians, who stopped feeling at home in an increasingly hostile province. They started to move out, taking the mining equipment back to the Soviet Union. Seeing falling revenues and Chiang Kaishek losing out to the Japanese, Sheng Shicai tried to change sides again, arresting Kuomintang officials on his territory. The move turned out to be a mistake which cost him power. Chiang Kaishek was still holding up, and in 1944 replaced Sheng Shicai with his appointee. It quickly turned out, though, that Sheng Shicai's ruthlessness, together with Soviet support, was the only thing holding the integrity of Xinjiang province together. Once these were gone, the Uyghur uprising broke out. With turmoil and lack of infrastructure connecting Xinjiang to inland China, profits from mineral resources were compromised. The Soviets took advantage of weak Chinese control and resumed illegal mining operations in sites closest to the border. By 1947, Soviets had been engaged in rogue mining and shipped what amounted to around 1,000 metric tons of lithium, beryllium and tungsten ore across

the border, mostly by the Irtysh River. The exterritorial operations were protected by the military with heavy machine guns.[63]

By the end of 1949, the Communists won the civil war, and
Mao Zedong established the People's Republic of China. The
attitudes toward the Soviets were warming, and besides, Mao
knew that he would not be able to make use of Xinjiang's wealth
without Soviet help. In 1950, two joint companies were established to commercialize Xinjiang's mineral wealth—the Sino-
Soviet Oil Company and the Sino-Soviet Non-Ferrous and Rare
Metals Company.[64] We are more interested in the latter, which
was occupied mainly with mining for lithium, beryllium and
tantalum. The Chinese real participation in those operations was
minimal. The capital, expertise and human resources yet again
came from Russia. The extracted metals were not used in
Chinese industry, but exported to the Soviet Union in their rawest form. Koktokay mines were a centre of activity. Its spodumene pegmatite reserves contained lithium and beryllium. Over
11,000 metric tons of beryllium and over 4,000 metric tons of
lithium were shipped to the Soviet Union between 1950 and
1954 from Koktokay. Minor metals such as lithium, beryllium,
niobium and tantalum were of crucial importance for the Soviets,
as they could not be found in sufficient quantities on their own
territory. Soviet investments in the Sino-Soviet Non-Ferrous and
Rare Metals Company grew explosively each year. But to the
surprise of their personnel, in 1954, and during deteriorating
Khrushchev–Mao relations, the companies were turned into
solely Chinese-owned entities. The Soviets were publicly thanked
for their effort. But in reality, nothing changed: Soviet experts
continued to work at the enterprise, and raw materials continued
to be shipped to Russian state-owned factories, often as a means
of loan repayments.[65]

In the 1950s China was focused on growing its heavy-industry
base. Central planners, however, quickly discovered that they

were lacking raw materials—the building blocks needed to transform a mostly agrarian society into an industrial one. Whatever they had internally was often in insufficient amounts, or of inferior quality. Minor metals were of particular importance for the cutting-edge technologies of the day. The Koktokay mine became central to the Chinese industrialization effort. China's increasing mineral resource-related activities in Xinjiang went in parallel with political efforts to align the region's power base closer with Beijing. To render the attitudes of the Uyghur population favourable towards the CCP, Xinjiang province was renamed as Xinjiang Uyghur Autonomous Region in 1955. Ethnically Han labourers and security apparatus have been repatriated to the region.

Truth be told, due to the underdevelopment of Chinese industry in the 50s, China had little use for minerals used in more sophisticated technologies, such as lithium. But lithium and other minor metals were instrumental in paying back Soviet loans. To expand their production, China decided to turn the Koktokay underground tunnel mine into an open pit mine. Thanks to that, in 1956, 16,600 tons of lithium ore was exported to the Soviet Union, double the amount of the previous year.[66] The Koktokay mine also changed the focus of production from beryllium to lithium. Between 1950 and 1962, it exported 100,000 tons of lithium and only 34,000 tons of beryllium to the Soviet Union.[67] At that time Koktokay was the only lithium mine in China. By the end of the 50s and the beginning of the 60s, lithium was mainly used in glass and ceramics production. Seventy per cent of its use in glass was in the rapidly developing television and electronic industries. The television boom in the USSR fell in the same period. Koktokay's lithium was crucial in making sure that every Russian family received their very own TV set. In 1958, '115 factory' was opened in Urumqi (Xinjiang), China's first lithium chemical factory, which processed spodumene ore inside Chinese territory into more added-value products such as lithium oxides and lithium salts—products which can be used not only in high tech but also in nuclear weapon applications. The lithium industry fuelled the development of highways and warehouse facilities in Xinjiang. The establishment of the Lanzhou–Xinjiang railway in the early 60s, inspired by the need to transport oil from Xinjiang to the interior, paved the way to future metals transports.

The campaign of the Great Leap Forward with its ultimate goal of the industrialization of the countryside served Xinjiang Non-Ferrous Metals Company well—by 1960, it had increased its workforce six-fold, to 24,000.[68]

The stream of USSR loans started to dwindle in the early 60s. The schism between the USSR and China took place on both ideological and pragmatic grounds. It stemmed from different interpretations of Marxist-Leninist doctrine, but also from the USSR moving away from the personality cult of Lenin and Stalin, which Mao had perceived as threatening. Meanwhile the Soviets' increasingly close ties with India were undermining China's security in the region. At that time, China was also invested in the development of nuclear weapons. The effort to 'build a bomb' in China started in 1955, after the Taiwan Strait crisis, when the US for the first time assured Taiwan of defence from communist invasion and even threatened to use atomic weapons as a means to that end.

China's first atomic facilities were built in Lanzhou, where the Xinjiang railway ended, and in Baotou, also known for its rare earth deposits. Lithium has been instrumental to China's nuclear plan. The lithium isotope lithium-6 is a fundamental raw material for the construction of atomic bombs. It reacts with a neutron to produce tritium—a key thermonuclear material for weapons. When tritium fuses with deuterium, it releases large amounts of energy, driving a thermonuclear explosion.

The importance of lithium in the nuclear race was reflected in production quotas at the Koktokay mine, when, in 1963, the output was officially shifted from prioritizing beryllium, used to pay the USSR's loans, to lithium, key in China's nuclear programme. Even nowadays, the progress of North Korea's nuclear programme has been gauged based on the completion of lithium-6 producing facilities near Hamhung in 2016, and Kim's regime business idea to sell excess lithium-6 online on global markets.[69]

When the criticality of Xinjiang's resources towards China's nuclear effort was identified, coupled with the importance of the region as an oil production base, the security apparatus and deployment of military personnel was increased throughout the 60s. Beijing still considered the scenario that the Soviets might have an appetite to annex the northern part of the province, where most of the raw materials assets were situated. The number of ethically Han Chinese in Xinjiang has grown, too, as mining towns expanded. Both increased security and the changing ethnic balance may have catalysed a growth of Uyghur–Chinese tensions.

The following years, which preceded the battery boom, were only mildly interesting, as far as lithium is concerned. The development of the lithium industry was driven by its applications in glass, ceramics, aluminium and air-conditioning industries.

Additionally to Xinjiang, two other lithium producing regions entered the scene, where there was a high enough lithium occurrence in the ground to justify its mining—Jiangxi, a poor, landlocked province with some proximity to the coastal regions, and Sichuan, a province bordering Tibet. The state-owned mines and processing facilities developed there, giving rise to today's two major lithium producers, controlling a large part of the global market share: Ganfeng Lithium, based in Jiangxi, and Tianqi Lithium, based in Sichuan. We will follow their story to global ascendency in the next chapter.

2

STRUGGLE FOR GLOBAL DOMINANCE

China arrived on the scene too late to compete with the established powers for dominance in the fossil-fuel-propelled world. However, the ongoing transition to electro-mobility powered by energy generated from renewable sources presents China with a great opportunity: to lead in a new industry, which the Chinese are calling the New Energy Economy.

The high GDP growth rate that Chinese society is so used to is becoming more and more difficult to maintain, especially when structural market weaknesses—impossible to avoid in a developing economy of that size—slowly start to resurface. But taking up this opportunity would create companies and jobs, and give the growth-hungry country some much-needed forward momentum.

There is yet another benefit. To keep its citizens content with the political status quo, the Chinese administration needs not only to maintain economic growth, but also to address—and eliminate—pollution. These seem like contradictory objectives in this highly industrialized nation. However, focusing development on the New Energy Economy—everything from the manufacture of pollution-free electric vehicles to dealing in the raw materials that batteries are made of—would address both these goals perfectly.

It is to this end that the party does what it can to erase road-blocks on China's path to an electric future—be it through tax breaks and subsidies, if consumer-driven demand does not meet annual targets, or through provision of cheap finance for acquisitions abroad, if the availability and quality of domestic battery metal resources raise concern.

But now, let us move away from this macro view and the motives to be found in Beijing's corridors of power, and look at the remarkable quest of individual Chinese companies: to ensure a strong resource base for the transition to the New Energy Economy by securing access to as much high-quality lithium as possible.

Jiangxi province is one of the poorest regions in China. In 2016, it was ranked as having the lowest wages in the country and the third lowest property prices; it is an unlikely place to start the story of a global conquest. Yet this is where Ganfeng Lithium, one of the biggest lithium producing companies in the world, known for its series of daring acquisitions in countries as diverse as Ireland, Mexico and Australia, was born.[1]

The company's founder, Li Liangbin, is as unassuming as the province he hails from. With a cardigan under his blazer, a prominent forehead and frameless glasses, he looks more like an academic than the chairman of a company offering a product associated with the new superpower's future. His estimated net worth of $1.4 billion in 2020 placed him 265th on China's Forbes list. This is not bad for a captain of an industry that continues to experience explosive growth, especially considering that Mr Liangbin still held around 20 per cent of Ganfeng Lithium's shares according to the 2018 annual report, making him the company's largest shareholder.[2]

Liangbin knows the lithium business inside out. After graduating as a chemical engineer, he went to work for the Jiangxi Salt Factory near his hometown, where he made a career switching between the roles of technician, engineer and researcher. According to his co-workers at the time, he was outstanding both in research and development (R&D) roles requiring a solid scientific background, and managerial roles, where soft skills and good relationships with people mattered. Liangbin's former co-workers are the first to claim that the company owes a lot to him, in terms of the technology and products he developed during his time there.

Eventually, his good work paid off and he made it to director of the company's lithium bromide department.[3] There, he worked with the air conditioning industry, where lithium bromide compounds are mainly used. He quit the company in 1997, right before the demand for lithium generated by lithium-ion batteries really took off in the early 2000s.[4] Perhaps he did not want to be stuck with air conditioners, when he saw a new important application for lithium approaching on the market. He took the difficult decision to leave a well-paid job in an established company and take a plunge into the entrepreneurial unknown.

The company that became the Ganfeng Lithium we know today started with the acquisition of a small lithium metal smelter from the Xinyu city government for what was then roughly $120,000.[5] From its humble beginnings to its listing on the stock exchange, Ganfeng was very much a family company. In its early years there were more family members among the company's workshop employees than outsiders.[6] Li Liangbin himself and his brother were supporting the fledgling enterprise with their own savings. Later on, as the business started to take off, more family members began to hold stakes and have roles in the company. Right before the stock exchange listing, Ganfeng's major shareholders were still Liangbin's brother, cousin, wife's brother and mother-in-law.[7]

Mr Liangbin was probably well aware that purely family businesses are not likely to grow into large organizations with a global reach, and the lithium industry, even in the early 2000s, was already a global business. So, in 2006, Liangbin expanded the company's ownership to include outsiders with the required skill set. Wang Xiaoshen came on board with not only excellent knowledge of the lithium market but, most importantly, fluent English.[8]

Both Mr Wang and Mr Liangbin started their respective lithium careers with state-owned enterprises. Mr Wang joined Xinjiang Non-Ferrous two years before Mr Liangbin started his first job. At that time, some graduates in fields key to the country's industrial development were being assigned to state-owned companies in need of personnel, without having much say in the matter. They couldn't freely choose where to start their career. Young Liangbin had a choice of either joining a steel mill with 50,000 employees or a lithium producer with 800.[9] His quantitatively oriented mind drew the conclusion that it would be easier to compete and make a mark in a smaller organization.

We can only speculate whether lithium's energy storage applications were on Liangbin's mind at the time when he started his own venture, but even at the very beginning lithium compounds used in batteries were a high-volume business. This was not to the extent of other key commodities such as grain, coal or oil, where most international transactions are made in whole-ship quantities, but it was still large enough to discuss deals in terms of full container loads.

However, the new company was too short on capital to deal in high-volume business.[10] Its founder had chosen a niche chemical over steel when starting his career. When launching his own business he fundamentally followed the same reasoning, focusing on more specialized lithium products such as lithium chloride and lithium metal to generate a cash flow.[11] Considering how small the lithium industry was at the time, it might be said he chose a niche within a niche.

Even at that time, his strategies proved innovative. Back then, lithium chloride producers relied solely on domestically produced lithium feedstock. Liangbin was the first to think about minting a relationship with the Chilean producer SQM—one of the largest producers in the world—to import the feedstock from Latin America, without intermediaries, and thus lower his own production costs.[12]

At that time, SQM was hardly a match for a small dabbler from China. The fact that Augusto Pinochet's son-in-law had a seat on its board spoke to the weight of this large chemical company, which then was better known for its potassium nitrate (one of the key ingredients in the growing premium fertilizer market) production than its lithium.

Nevertheless, Liangbin somehow managed to convince the Chileans to work with him. Now, take a moment and put yourself into the shoes of a small lithium producer. You will quickly realize that one of your main problems will be to get cash quickly enough to pay for your inputs, so that your expensive processing equipment does not stay idle. It takes around a month to get your input material from Chile, and usually you need to pay in advance to make the seller load a container worth over $100,000 on the shipping vessel. Then it takes further days, or even weeks, to process the material to get a marketable output. Once you have an output, you need to sell it, in an environment where demand for chemicals and commodities is cyclical. If you supply lithium compounds for use in consumer batteries, you may expect the demand to pick up in preparation for the Christmas period, when electronic devices containing them sell well, but the summer period might be flat. Lack of demand may force you to keep the stock in anticipation of future sales, while your plant needs to be running, and your bills and employees need to be paid. Just imagine the anguish of dealing with cash flow under those circumstances. Even if you are profitable on paper, you can still go bust.

Amid all these hazards, Liangbin was even keen to offer a 25 per cent stake in his company for free to SQM, just to secure worry-free access to the essential raw materials for his plant.[13] Reportedly, China's SQM representative was so shocked at the idea that he told Liangbin it was too much of the company to give away for free—plainly forgetting for a moment whose interests he was representing.[14]

On the flight to Santiago, Liangbin might have pondered different scenarios in his head. In the end he took the SQM's representative words to heart, and proposed a lower stake of 15 per cent in his company for free, in exchange for support in the supply of feedstock.[15] SQM was interested in the deal—enough to start the due diligence process on Ganfeng. Yet after checking the company and assessing its prospects, the Chileans rejected the proposal.

It took only two years for SQM to change their mind. They came back with an unsolicited bid, offering a competitive price for shares. But by then Ganfeng was strong enough and determined enough to make it on its own, and declined.[16] Another twelve years later, the companies would change places: it would be Ganfeng bidding for SQM's stake in a promising early-stage Argentinian lithium mine project.[17] This time, the deal would be concluded.

There is more nuance to this story of the relationship between SQM and Ganfeng. It was Wang Xiaoshen, the first outsider to join Liangbin's company's inner circles, who distributed SQM's lithium products in China in the early days, while still working for Xinjiang Non-Ferrous.[18] And it was probably through Xiaoshen that Liangbin learned about the advantages of SQM's offer and, albeit indirectly, established the first links between himself and the Chilean player. As much as Liangbin took the company off the ground, it was Xiaoshen who proved instrumental in establishing Ganfeng's international presence and leading its chain of foreign acquisitions. Often taking a seat on the board of acquired companies, Xiaoshen's experience and English-language skills made him the face of Ganfeng and its window to the world. Yet for this to happen, the organization needed to acquire more funds to turbocharge its growth.

Markets, particularly growing markets, do not tolerate a vacuum. In new industries, where demand is strong, a 'winner takes all' mentality prevails. This is reflected in the story of Rockefeller's Standard Oil—one of those stories that shaped the vision of corporate America. As hard as it might be to imagine, in the second half of the nineteenth century—the early days of Rockefeller—oil was a nascent industry. Its main application at that time was not that which ultimately drove its growth; back then, it was mainly used in the form of kerosene, to provide a source of better, less polluting lighting. Similarly, back in 1994, only 7 per cent of lithium produced in the world was used in battery production, while the major consumption drivers were applications in ceramics and glass.[19] Rockefeller quickly dominated the early oil industry through vertical integration, making investments and acquisitions across the value chain that allowed Standard Oil to streamline production and logistics, lower costs, and therefore undercut competitors.

The strategy of the major Chinese lithium players—not only Ganfeng but also Tianqi's Lithium, which we will soon discuss—are not much different. Ganfeng's LinkedIn entry reads: 'We have a vertically integrated business model, with operations along the critical stages of the value chain, including upstream lithium extraction, midstream lithium compounds and metals processing as well as downstream lithium battery production and recycling'.[20] The Ganfeng approach has always been, in the words of its chairman, to follow a step-by-step strategy and to set achievable goals.[21] It is this strategy that has led the company from being a local, family-run business to a major global player.

Yet, as noted above, in order to enter its global expansion stage, Ganfeng first needed an extra financial boost. Even though company revenues had grown organically, in the turbocharged battery economy, they needed leverage from capital markets to stay ahead. Investment in Ganfeng by an unlikely trio of strategic investors—China-Belgium Investment Fund, Nanchang Venture Capital and China's state-owned Minmetals, one of the largest metal companies in the world—was a major step forward towards the listing of Ganfeng on the stock exchange.[22]

Listing on the exchange not only opened up a new era for the company, it also made its founders and key executives multimillionaires.[23] Every private company that enters the stock exchange needs to issue a prospectus, a marketing tool and legal document that the company uses to market its shares to the public and communicate key information about the company and the IPO (initial public offering). Investors use the prospectus to help them make informed investment decisions. According to Ganfeng's prospectus, thirty-one of its thirty-four shareholders were private individuals, as opposed to other legal entities.[24] Liangbin's family altogether held 33,697,425 shares, a staggering 44.9 per cent of the total issued share capital.[25] The close circle who had been supporting Liangbin, on his difficult journey from a smallish regional smelting workshop to a company entering the lithium premier league, had their payday.

This evolution of Ganfeng took place at the right time, and arguably might be very difficult to recreate in today's China. This is partly because the lithium market subsequently took off and became more competitive, and partly for environmental reasons. Until 2013, China's government ignored all environmental issues and was very touchy if the rest of the world tried to broach the subject, even though China's pollution problems had major implications outside its borders. In February 2013, Japan's foreign ministry proposed a meeting with Chinese officials about the choking smog from Beijing that was becoming detectable on Japan's territory. Japan issued a statement that reverberated through the international press: 'The Japanese government is concerned and is closely monitoring the recent serious air pollution in China', arguing that China's pollution levels were affecting Japan's environment and the health of its citizens.[26]

The line taken by China's administration at that time was to accuse Japan and the rest of the world of hypocrisy. Beijing's mandarins argued that when the Western world and Japan had been rapidly industrializing, they had done so with no thought for the environment. So why should China, now that it was on the same trajectory, slow its pace in the name of sustainability? Yet ultimately, it was internal pressures, not external, that eventually led to China's sudden turnaround in policy. Smog in large cities, and especially in the capital, made it difficult to carry on for everybody. The winter season of 2012/13, when the pollution level in Beijing at one point hit forty times recommended safety levels, became China's environmental winter of discontent, providing a final push to trigger new policies. Chen Guangbiao, a businessman and philanthropist who made a fortune in recycling, started to sell cans of fresh air sourced from China's pollution-free regions. He sold ten million roughly Coca Cola-size cans in ten days, explaining: 'I want to tell mayors, county chiefs and heads of big companies: don't just chase GDP growth, don't chase the biggest profits at the expense of our children and grandchildren and at the cost of sacrificing our ecological environment'.[27] The success of this curious enterprise serves as great anecdotal evidence of how desperate Chinese society was to do something about the smothering smog levels.

If you are a leader of an authoritarian government, your main advantage is that as long as you stay in power you can think long-term. Your plans are executed and rolled out effectively from top to bottom without much hindrance. So when the

Chinese administration put its mind to finally tackling environmental problems, the impact became visible very quickly.

The war for clean air started on two fronts, affecting both households and industries. Households were banned from burning coal, and in certain areas inspectors physically removed coal boilers from people's homes and replaced them with electric or gas stoves. While waiting for replacements, some people suffered through long winter weeks without heating. School children had to be brought outside from stone-cold classrooms to school yards to get some warmth from the winter sun.[28]

Dramatic as this fast transition was, in our story we are more interested in the impact on industries—and here the pace of change was spectacular. Between 2013 and 2018, the smelters and mines across the metal spectrum—from niche metals like antimony to non-ferrous metals such as aluminium—that are traded in large volumes on global financial exchanges were affected. China is the largest producer of refined copper, lead, zinc, tin and aluminium. And a significant part of China's metals output has come from relatively small companies, very often operating, at least in part, illegally. All these facilities came under sudden scrutiny to make sure they met tightened emission standards—but with the low volumes at which they operated, they simply could not afford to deploy the technologies necessary to meet anti-pollution criteria. Many of these companies have gone out of business—and young Ganfeng Lithium could have met a similar fate had it started later than it did.

Nobody has really measured the extent to which China's new environmental measures have resulted in lost capacity across the metals spectrum, but for zinc alone, Platts, one of the top commodities market research firms, wrote: 'We estimate that China has reduced its share of global mine production from 40% in 2016 to 34% in 2018 and understand from market anecdotes that environmental standards are the dominant driver in these closures'.[29]

Overall, China's policy bore fruit. By the end of 2019, Beijing had exited the world's list of the top 200 polluted cities.[30] China's capital's PM 2.5 levels have been reduced by 35 per cent from 2015 to 2018. PM 2.5 stands for particulate matter, a term encompassing a diverse mixture of solid particles, including the output of fossil fuel combustion. Critics would point out that PM 2.5 levels are still roughly four times higher than those in London, but considering the short time span within which this improvement has been achieved, the results have been truly remarkable.

Back to Ganfeng. After the listing on the Shenzhen stock exchange, the company found favour with many small-time retail investors, upbeat about the new energy economy. In Xinyu, where Ganfeng is based, Liangbin's success galvanized the imagination. Everybody and their dog was buying Ganfeng's stocks.[31]

In an interview for the local Jiangxi News Network in 2018, Liangbin spoke about corporate governance, investor relations and further expansion abroad.[32] Those concepts are neatly related. Corporate governance and investor relations are the bread and butter of stock listed companies from the West. Yet in China, where family businesses or state-owned enterprises dominate, this is not the case. A proper corporate governance model, guaranteeing the stability and operational continuity of an enterprise as a standalone entity, delinked from its owners or high-level executives, is something to be proud of, especially in Jiangxi province business circles.

Ganfeng's expansion strategy seems to be trifold. Domestically, it is a strategy of mergers and acquisitions of companies complementing the existing supply chain. It is a 'big eats small' kind of strategy. On an international level, the strategy has two elements: making sure that an adequate amount of feedstock is secured for processing into battery-grade materials, and gaining access to investors outside of China. To achieve the latter, Liangbin set his sights on ringing a bell in Hong Kong's Victoria Harbor.[33]

A bell is physically rung upon the opening of stock exchanges around the world, and traditionally the privilege of doing so is given to a representative of a newly listed company on its first listing day. Listing on the domestic Chinese stock exchange in Shenzhen provided the company with access to capital from China's domestic investors, but Liangbin soon developed an appetite to tap international capital markets—and the first step to realizing this dream was to list in Hong Kong, the traditional financial gateway to China. Crucially, that could not happen without Ganfeng's strong corporate governance, which provided protection against insider trading—where people close to the company trade on information that has not been made public, thus getting an unfair advantage over other investors.

Surely fairness and transparency in respect of investors is a sensible metric for determining whether a domestic Chinese company is ready to tap into pockets beyond its borders. Yet it is less clear what impact the support from the Jiangxi Provincial Party Committee and the provincial government—which Liangbin mentioned in an interview (with China's Jiangxi Network in 2018) as being crucial for Jiangxi-based companies—had on the road to listing in Hong Kong.[34]

What Liangbin's comments to domestic media make clear is that for the company to grow, the access to upstream high-quality lithium resources globally needs to be secured.[35] The term 'upstream' has been adopted from the oil industry, and designates the part of the supply chain closest to the rawest form of the commodity. Lithium is generally traded in the form of ore, concentrate and lithium compounds. In its crudest form, lithium is mined as ore. But what is interesting about ore is that the amount of lithium contained in it varies, even if it comes from the very same deposit. Variability is, however, within a very narrow and small range of 1 to 2 per cent of lithia in the ore.[36] The rest of the ore, so a minimum 98 per cent of its weight, is composed of other minerals, usually of no interest to lithium miners or just waste with no interest to anybody at all. The direct implication of this situation is that economically it does not make sense to transport lithium ore over large distances unless there is absolutely no way to enrich it nearby.

If we look at Australia, currently the world's largest producer of unrefined lithium, for some years after it became involved in the lithium mining business it mainly exported the product to China in ore form, as there were not enough processing facilities within the country.[37] Generally, the mining step, be it for hard rock deposits or for salt lake brine deposits, is the easiest processing step—although this does not necessarily mean that it is an easy step. It is the further processing that is more difficult, and the further we move along the supply chain, in the direction of the material used in batteries, the more difficult processing becomes from an engineering perspective. As soon as the concentrating facilities were built, however, Australian producers immediately ceased exporting ore to China and started to export lithium concentrate instead. As the name reveals, the amount of lithium in concentrate is 'concentrated', usually to the level of 6 per cent of lithia.[38] It may still seem not very much, but it is at least two times the lithia content found in ore, so from a cost perspective it absolutely makes sense.

With this in mind, you may well ask why lithium in Australia is not refined even further. It does seem as if this is the direction in which Australian producers are going, but this road leads either uphill or through swamps. Even though the development of converting capacity, which takes lithium concentrate from 6 per cent lithia to lithium carbonate battery-grade material with a minimum of 99.5 per cent lithium content, takes much less time than the development of a mining operation, there is still a huge amount of know-how and experience required for this chemical step. It is also not just about raising the amount of lithium content in the final product, but about keeping the impurity levels (magnesium, potassium, sodium) under control, as these in the end can affect the performance of the cathode—the battery's key component.

It is also not a zero-sum game, where you either get a pure battery-grade compound that EV makers would kill for, or good-for-nothing industrial waste. If you build conversion capacity, you may as well produce a technical-grade compound that you can sell to glass or ceramic makers—or to your Chinese partner who has the converter and engineers with the know-how to take it a notch higher and further purify it to battery grade.

Ganfeng, we managed to establish, is after quality raw material lithium, to buy and further process into the material that can be sold to EV battery components makers. Whether the availability they have in mind has any temporal connotations or not is harder to say. The question is whether they think there is a limited quantity of high-quality lithium resources in the world in general, or whether they think there is a limited amount of accessible lithium resources that can be mined over the next five to ten years—when EV sales are supposed to really take off. Because if you do not carve out the largest possible market share for yourself as a producer in good time, then it may become increasingly difficult for you to do so later on. In this business, as in any industry that is new and experiences rapid growth, it is the early bird that catches the worm.

Ganfeng's continuous expansion plans put it on the way to becoming the world's number one lithium producer. Yet some of the company's critics and competitors say that this title does not matter. The current number one, Albemarle,[39] has declared its intention of being the most profitable company in the lithium sector.[40] It wants to create as much value as it possibly can for its shareholders, and does not seem to care so much about being 'number one' in terms of pure volume. Albemarle says what stock investors love to hear. And thinking like this—in terms of profit—is absolutely healthy for stock-listed American companies—but it misses a more strategic angle, one which instead transpires in the actions of China's industry enterprises.

As those companies understand very well, their thinking needs to be aligned with the strategies of the authorities in Beijing, who want supply chain security and cheap raw materials to make transport electrification happen. In a country as large as China, this can only be achieved by developing lithium and battery material industries at scale. China has travelled this road already with steel, aluminium and solar panels.

Let us have a closer look at the development of the aluminium sector in China. As is the case with lithium, the automotive industry has been and continues to be the main demand driver for this metal. In 2005, China's share of global aluminium production was below 10 per cent, yet it grew to 57 per cent in 2018.[41] Does this extreme growth in capacity mean that the trend for aluminium has been positive through all these years? No; in fact its prices have experienced a lot of volatility over the last fifteen years. Between 2004 and 2007, during the global commodity boom—driven by China's extraordinary annual GDP growth rate between 10 and 14 per cent—aluminium prices grew steadily from \$1,715/ton to \$2,638/ton.[42] China's production capacity at that time grew as well.[43] But in 2007 the tide changed, and between then and 2009 prices fell, magnified by the global recession following the 2008 financial crisis. China's production decreased very slightly at the time, but its share of the world's output expanded. After 2009, aluminium prices rebounded again to \$2,400/ton, with a substantial growth in China's production levels and world market share.[44] Then the years 2010 to 2016 brought yet another long period of price decline. So did this rollercoaster hinder China's production expansion at any time? No; aluminium output growth was unaffected and has been so large that China's world market share has grown from around 10 per cent to over 60 per cent.[45]

Most of the 'good years' in the aluminium industry have been driven by China's hunger for resources and the import of aluminium from abroad. Aluminium producers outside of China operated in negative margin territory for long periods of time, as they hoped in vain that China's generated demand would finally wipe out market surpluses. As those hopes faded away, they were forced to cut back on capacity. Meanwhile, China had been expanding its capacities continuously, irrespective of the price situation. This was because the overarching rationale for growth was the independence of China's supply chain and reducing the cost of a key raw material for China's automotive industry—and not the maximization of shareholder value.

So where are we now, in the lithium market, to draw a parallel with the rollercoaster of aluminium prices? At the end of 2019, lithium prices experienced a drop of over 50 per cent on average.[46] I say on average, as lithium is being traded at different stages of processing and in different qualities, on different types of contracts which differ vastly in sensitivity towards changes in market environment. Saying that, the general trend has been downhill. There was also no significant uptick through most of 2020. Only at the very end of the year, lithium carbonate prices started to move up in China.[47]

One of the most successful investors of our times and a boogeyman of the far right, George Soros, says in his reflexivity theory of markets that an objective view of the market is impossible to achieve for individuals, hence we should always assume that our view is biased and incomplete. Further, this investor's bias, be it negative or positive, has an impact on the market and becomes a self-fulfilling prophecy.

Starting around 2011, analysts forecasted a dramatic capacity expansion in lithium mining operations, which was supposed to become operational around 2017. When the year finally arrived, over half of the predicted capacity did not materialize, while EV growth began to take off at an impressive pace, especially in China. This made market participants panic that soon there would not be enough lithium to enable further EV adoption. Prices went through the roof, to levels unknown to the industry's veterans. In June 2017, the *Financial Times* published an article under a title reflecting the frenzy: 'Electric car demand sparks lithium supply fears'. It quoted John Kanellitsas, vice-chairman of Lithium Americas, who said: 'There's much more consensus on demand; we're no longer even debating demand. We're shifting to supply and whether, as an industry, we can deliver.'[48] Guess what—at the end of 2019, with lithium oversupply on the market, people were debating the demand, at least in the short term. The forecast horizon line has moved for industry bulls, to 2025, where a significant deficit is likely to appear according to analysts' consensus.

Many types of organizations, from investment banks to lithium producers to government agencies, employ analysts who try to forecast where the price will be in the short, medium and long term, together with demand and supply. Most companies tend to stick to similar views and numbers, what is broadly called analysts' consensus. Breaking away from the analysts' consensus is a risky endeavour, provoking fierce attacks from other analysts on Twitter or during presentations at industry conferences. Thus, for analysts, it is safer to move in herds, offering forecasts on uncertain futures, with a very limited variance.

Since EVs are the main driver behind lithium consumption— cell phones contain around 5g[49] of the element while fully battery powered EVs 30–60kg[50]—it all comes down to the question of whether people will go for electric vehicles, and when. Sure, there is a strong regulatory push involved, to make life greener for all of us and to stop climate change. But in the end, it comes down to you and me, and whether we choose electric over gasoline powered cars.

Rational arguments behind buying or not buying involve the cost of electricity over the cost of fuel, driving range on a single charge, availability of charging points, safety of the battery and how long it takes to charge up. And psychology, often omitted in EV market analysts' disputes, also plays an immense role in consumer choices. So as long as we talk about massive, market altering adoption, it is as much about making EVs cheap and efficient as it is about making them sexy. Otherwise, EVs will be in danger of remaining a niche product on the verge of conventional, mass-mobility solutions.

Ganfeng currently holds stakes in seven[51] lithium resources around the world, including spodumene ores, lithium-containing brines, and clays.[52] Nowadays the majority of lithium is produced from spodumene ores, but for decades this was not the case. Spodumene ore was originally applied in the glass-making and ceramic industries.[53] In the glass-making process it was employed to lower the melting point[54] and hence save energy and thus costs during production, or to produce high-strength glass.[55] In the ceramic industry, it was used to produce the composite of glass and ceramics that you can find as the typically black surface on your induction cooktop. But technology improvements and demand pressures turned spodumene (hard rock) mining upside down, and the whole converting industry was born on a large scale. The vast majority of spodumene converting facilities are based in China. Some of these are independent enterprises, built by businessmen experienced in vastly different industries in search of new profit opportunities. A lot of those new companies started with the idea of catering to the sprawling EV industry, but ended up selling material of a lower quality, more likely to find applications in electric bikes or the electronic equipment of lesser known brands. Production of the higher quality material turned out to be much more difficult than expected.

Lithium trade between Australia and China, as we stand, is reminiscent of the trade in iron ore. Large quantities of lithium are shipped, mainly from Western Australia's Port of Hedland, as the biggest lithium mines are localized in its vicinity.[56] The difference between iron ore and the lithium trade is that iron ore is shipped in cruder form than lithium and in much higher volumes. It is all related to the incomparably larger scale of the steel industry. We need to remember that steel production was behind China's industrialization, but that it also drove up the pollution levels. Now we find lithium as a material behind another major change, where electrification, decarbonization and digitalization are the names of the game.

It is fascinating to observe how Australia has managed to position itself to profit from China's transitions, providing the raw materials to make them happen. In 2018, iron ore accounted for a stunning 15 per cent of the value of all Australian exports.[57] Lithium is unlikely to achieve this scale anytime soon. When we look at the bigger picture, however, it is not only about the size of the revenue but about strategic positioning. For the foreseeable future, China will need Australian lithium resources to continue the phasing out of internal combustion engines. This situation will provide Australia with a unique negotiating position in its relations with China.

As a trade war between the US and China emerged, comments resurfaced as to how China could weaponize its rare earths deposits. Rare earths are minerals used in a number of cutting-edge technologies, particularly in the production of permanent magnets with key, irreplaceable applications in defence, high tech electronics and EV motors. In May 2019, the editor in chief of the *Global Times*, considered to be the Chinese Communist Party's mouthpiece on international affairs, said on Twitter: 'Based on what I know, China is seriously considering restricting rare earth exports to the US. China may also take other countermeasures in the future.'[58] China controls over 80 per cent of the world's rare earths production, and has pointed a metaphorical gun at the US and Japan on a number of occasions when relations between the countries deteriorated.[59] In the case of Japan, it once pulled a trigger. On the morning of 7 September 2010, the Chinese trawler Minjinyu 5179, operating in the East China Sea near the disputed Japanese Senkaku Islands, to which China stakes territorial claims, collided with the Japanese Coast Guard's patrol boats. The collision ended in the detention of the Chinese skipper. China, after making demands to release him for several days, eventually withdrew from the talks. Meanwhile it applied another, much more effective, type of pressure: stopping the exports of rare earths to Japanese companies. Even if there was never an official embargo, Chinese custom officers stopped rare earths shipments to Japan for additional inspections. The official rationale that was given was the need for counter-smuggling checks. The supply chain bottleneck lasted until some time after the skipper was released, and affected the markets beyond Japan, raising rare earth prices globally. This was to the detriment of some of Japan's most innovative companies. Shortly after, the Japanese trade minister himself pleaded with China to restart the exports. Soon the exports normalized, and no lasting mark was left on the economy. But concerns remain to this day, both at the managerial and the political level, that China might wave its 'rare earth gun' again.

Japan occupies a very sensitive place in the global high-tech supply chain—it is situated exactly in the middle. It buys raw materials from China and transforms them into building blocks for the electronics industry, in the form of components or advanced chemicals. Japanese companies managed to develop and patent these products during Japan's economic miracle in the 80s and early 90s. Later on, when South Korea started its rapid economic ascent—driven to a large extent by the production of electronic devices, especially mobile phones—its domestic companies did not have to reinvent the wheel. They were building pretty much on top of the fundamental technologies provided by Japan. South Korea became acutely aware of this dependency only recently, during the Japan–South Korea trade dispute, which we will look into later in the book.

China's embargo on exports of rare earths to Japan was not even difficult from a legal perspective. China's largest rare earths producers, such as Chinalco, Northern Rare Earth, Xiamen Tungsten and China Minmetals, are all state-owned, and the production and export of rare earths is regulated by quotas even under regular market conditions.[60] It is not unreasonable to speculate that if Australia continues to ramp up its lithium feedstock production, and China does not manage to further expand its domestic production of lithium's raw materials, the day may come when it will be Australia that will have an interesting point of leverage in the trade disputes of the future.

Chinese lithium resources are large, especially the salt brines situated on the Tibetan plateau. But this location makes for a harsh environment to operate in and transport the cargo, and, to make it worse, Tibetan resources are characterized by a high level of impurities, particularly magnesium.[61] As we discussed earlier, producing good quality battery-grade material is mainly a matter of keeping impurities under the cap. Plus, the Tibetan Autonomous Region, despite billions invested by the Chinese and a tight security apparatus, is not exactly the embodiment of political stability. Realizing these deficiencies of available domestic resources, Ganfeng therefore acquired stakes in Mt Marion and Pilgangoora—Australian hard rock spodumene mines. Ganfeng, however, does not wield full control of these assets.[62] Even if it did, the output might still be subject to Australian administration-controlled export tariffs, royalties and production or export quotas. To point out the obvious, even full ownership of an asset in foreign jurisdiction does not allow the overseas owners full control over the material. Mt Marion and Pilgangoora are Australia's second and third largest lithium mines. The share that Ganfeng hold in Pilgangoora is small—4.3 per cent—while in Mt Marion it reaches a highly significant 43.1 per cent.[63] But even a small percentage share is enough to be an offtake partner of choice.

The largest lithium mine in Australia is Greenbushes.[64] With over 100,000 tons LCE (lithium carbonate equivalent)[65] of spodumene concentrate produced in 2019, it is the largest by a wide margin, considering that a remaining 140,000 tons LCE of the production is split among another six mines.[66] Greenbushes mine is also to some extent Chinese, with 51 per cent belonging to Ganfeng's archcompetitor, Tianqi Lithium.[67] At this stage in our story about lithium, it is time to get to know another important Chinese player a little better. The competition—or rather a friendly rivalry—between the two players takes place not only on domestic, but also on global ground. Many parallels can be drawn between Tianqi and Ganfeng. Tianqi, in spite of being a listed company with a market cap in billions of dollars and a global footprint, maintains the character of a family-run[68] company.[69] Compared to Ganfeng, however, it has a more politicized founder at the helm.[70] Mr Jiang Weiping's education and a large part of his career was bound up with agricultural machinery.[71] Like Ganfeng's founder, he fell into the lithium business by seeing an opportunity others had not. Mr Jiang started out by buying an ailing lithium compounds production business from the state—Shehong Lithium.[72] Together with his family he now holds more than 41 per cent of Tianqi Lithium's equity, mostly through control of the entity Tianqi Industrial.[73] Jiang Weiping holds 88.6 per cent of Tianqi Industrial, while his daughter Jiang Anqi holds 10 per cent.[74] As director and general manager, Ms Yang Qing holds the remaining 1.4 per cent.[75] In addition, the founder's wife is Tianqi Lithium's second largest direct share-holder, holding 5.16 per cent of the company.[76] Keeping owner-ship in the family has made Mr Jiang very rich indeed. China's Hurun list—a local equivalent of the Forbes list—estimated his wealth at around $1.8 billion in 2019.[77] He also dabbles quite successfully in politics. In 2018, he was elected as the representa-tive of Sichuan province to the National People's Congress, the highest organ of state power and the national legislature of the People's Republic of China.[78] In a number of interviews, not surprisingly mostly with Chinese media, Mr Jiang underlined his enthusiasm for manufacturing, relating how he likes to listen to the rumbling sounds of the machines on his factory floor.[79] Lithium, the third element in the periodic table, the most active and the least dense metal, had grabbed his interest due to its unique properties and wide ranging applications, from nuclear industry to batteries. He facilitated lithium bearing spodumene sales to China, from Australia, when the local producers failed to see its growing importance.[80]

Back at a time when lithium prices seemed to only go in one direction—up—and everybody was concerned with an imminent supply shortage, Mr Jiang stayed cool. He lived through the ups and downs on the lithium market, and warned about overcapacity build-up at the low end.[81] For him, lithium was a cyclical busi-ness, which sooner or later would show who was swimming naked when the tide of low prices finally arrived.[82] To counter that, he needed to secure production of high-quality material at a low price. Tianqi lithium's strategy reflected his views. The company has been very expansionist;[83] its founder did not worry about demand not materializing, but looked to secure good qual-ity assets on the mining side and to ramp up chemical compound production. And the behemoth Greenbushes is not the only highly prized jewel in Tianqi's crown. Perhaps even of greater importance to the company is a stake in one of the largest lith-ium companies in the world and a direct competitor: SQM.

SQM, unlike Greenbushes, is not just a mine. It is a vertically integrated 'mine to battery materials' producer. Tianqi acquired a 23.8 per cent share in SQM from the Canadian fertilizer company Nutrien for $4.1 billion, making it the largest deal in the history of the lithium industry.[84] This deal came about among many controversies. The story starts with a seller, Nutrien, a fertilizer giant and the world's largest potash producer.[85] Potash is one of the main ingredients of fertilizers, and Nutrien was created when two companies, Agrium and Potash Corp, merged on a wave of consolidation in the depressed fertilizer market. Fertilizer producers witnessed a rapidly growing global population that needed to be fed, and rising middle classes that wanted to eat more protein from livestock—which needed to be fed a grain-based diet. Following this train of thought, they believed that the demand for fertilizers would skyrocket and invested in a capacity build up that soon turned out to greatly exceed actual needs. The merger had been approved by most of the regulators in the countries where the two companies had been active. Only Chinese and Indian watchdogs had raised a red flag regarding this deal, seeing in it the potential for a monopoly.[86] China and India are two of the largest consumers of fertilizer in the world. Losing access to any of these markets would be a disaster for a newly merged company. In the end, an agreement was reached that the merger would be approved if Potash Corp were to sell its 32 per cent stake in SQM, which, alongside its lithium interests, was also a large fertilizer player.[87] In fact, in 2018, SQM's sales of potash and fertilizers constituted 47 per cent of its revenues, while lithium, thanks to which the company made frequent headlines in the media, made up only 32 per cent.[88]

Potash Corp agreed and the merger between it and Agrium went ahead to form Nutrien. Soon after, an investment bank was hired to help auction its SQM stocks. The deadline set by the regulator to finalize the sale was March 2019. Out of

32 per cent of SQM that had to be sold under the agreement, 8 per cent was auctioned publicly on the Santiago Stock Exchange, collecting $1 billion.[89]

It is unclear who approached whom first regarding the deal between Potash Corp and Tianqi. Had Tianqi had an eye on the expected sale for some time, or was it the bankers who gave Tianqi a call? In retrospect, it does not matter much. What does matter is that the deal was finalized as the largest ever in Chilean stock market history. But Tianqi was not financially ready to act on this once-in-a-lifetime opportunity. They needed a loan from the Hong Kong-based CITIC Bank in order to finance the SQM shares acquisition.[90]

CITIC is one of the largest banks in China, created by the so-called red capitalist, Rong Yiren, who was also once China's vice president.[91] The nickname comes from the fact that his family was one of the very few of the pre-1949 Chinese bourgeoisie who managed to find a way to work with the communists. His party membership had been kept a secret, revealed only after his death.

The deal had been controversial in Chile from the very beginning, with different actors trying to block it. The first opponent was Julio Ponce Lerou, éminence grise of the lithium world, and Pinochet's erstwhile son-in-law. The acquisition was originally set for fast-track antitrust approval, but Mr Ponce Lerou filed a claim before the constitutional court against the accelerated proceeding.[92]

Julio Ponce Lerou is SMQ's largest individual shareholder, with a 30 per cent stake, which brings his net worth to $4.3 billion according to Forbes.[93] Officially, Ponce Lerou and the organization through which he holds the shares—Pampa Group—complained that the acquisition would not promote free competition, since it would authorize Tianqi to participate in the ownership and management of another direct competitor.[94] The real reason why Mr Ponce Lerou opposed Tianqi, however, might be that the acquisition would create a counterweight within the company, limiting his ability to steer it. To Ponce Lerou's dismay, the constitutional court ruled in favour of the fast-track approval, stating that his claim lacked grounds.[95] It was not the end of legal headaches for Tianqi, though.

CORFO is the government agency created to promote economic growth in Chile, and it calls the shots in the domestic lithium industry through its control of the country's lithium leases. Its head at the time, Eduardo Bitran, had been very much involved in pushing for the development of the domestic lithium industry. Under his leadership, CORFO filed a forty-five-page complaint with the regulator, arguing, like Pampa Group, that the deal would endanger free competition on the lithium market.[96]

Members of Michelle Bachelet's government tended to be against the proposed Chinese acquisition, and they were not alone among political circles. A presidential candidate of the centre-left, Alejandro Guillier, filed a complaint as well.[97] Amid growing political pressure, FNE (the Chilean antitrust office) finally launched an investigation that identified several concerns.

Happily for Tianqi, the political climate then changed in Chile. Bachelet's administration, together with the head of Corfo, Eduardo Bitran, was entirely replaced when Sebastián Piñera took presidential office.[98] Tianqi finally managed to reach a settlement with FNE, committing to take a wide range of measures to comply with antitrust laws, including a ban on the election of board members, and restricted access to some of SQM's confidential commercial data.[99]

This was a big win for Tianqi, but soon after, many analysts put a question mark over the viability of the transaction, wondering whether Tianqi would have enough money to pay the acquisition loan back. To help with the upcoming payback, Tianqi was listed on the stock exchange but raised only $424m in December 2019—less than half of the amount they expected.[100]

The whopping $2.2bn repayment to CITIC bank was unlikely to make Tianqi go bust, however.[101] This is simply not how the system works. In China, there are around 150,000 state-owned companies.[102] They are not always profitable and they rarely receive what would be classified in Western economies as state aid, for instance grants. Yet at the same time, the state helps them even under enormous financial stress. For China's party officials, both on national and provincial levels, social stability is key. Nothing makes people go out on the street to protest like job losses. Most Chinese banks, including CITIC, are in fact state owned. CITIC bank is a fully owned subsidiary of China International Trust and Investment Corporation, which in turn is a fully state-owned investment company. To believe that a fully state-owned investment entity would allow one of the two largest companies in the strategic industry of the future to fail is a little naive. In fact, thousands of permanently unprofitable state-owned companies, in a range of industries characterized by over-capacity, are being supported by perpetually extended loans from state-owned banks.

The stereotype of a hard working, frugal Chinese society seems true to an extent, at least in regard to individuals. Levels of consumer loans and even mortgage loans are low in China by any Western standards. The same is true of loans to privately owned SMEs—yet the amount of debt amassed by large companies has skyrocketed in the last decade, at a growth level not found in more advanced economies.

The Chinese government is not exactly happy about this situation and wants to de-leverage the economy in the long term. In November 2020, the state-owned Yongcheng Coal and Electricity Holding, once a very successful coal mining operation, defaulted on its bonds, with the government failing to come to the rescue.[103] This changed investors' assumptions and raised borrowing costs for state-owned enterprises. Still, we need to keep in mind that Beijing allowed one of its many coal mining operations to go under, as the country transitions to cleaner sources of energy. This does not necessarily herald the emergence of a new trend that would affect China's major lithium player.

The most recent development, as of early 2021, is that Tianqi found a strategic investor in its Australian assets. IGO Ltd is an Australian company, with lots of operational experience in the country's mining sector. Lithium turned out to nicely complement its nickel-cobalt-copper mining operation in Western Australia, as it transforms itself into a significant player in clean energy metals. Tianqi entered the agreement to sell a substantial share in its Kwinana lithium hydroxide plant and Greenbushes lithium mine, to the tune of $1.4 billion, while maintaining control.[104] If finalized, the transaction, together with rebounding lithium prices, is likely to get Tianqi out of financial trouble, without a recourse to state aid. The involvement of the Australian IGO could also be a good move politically, as bilateral relations between Canberra and Beijing had severely deteriorated throughout 2020. Australian politicians took a stance on most of the issues seen as sensitive in China. At the same time, a document (most likely deliberately leaked) from the Chinese embassy listed fourteen grievances that the Chinese authorities hold against the Australian administration, including spearheading the crusade against China in multilateral fora and attempts to 'torpedo' its Belt and Road Initiative.[105]

The lithium industry is happy with the Chinese influence that largely characterizes it. Mining is always perceived as a large industry, with money. If you tell people at a cocktail party that you own a mine, by default you are assumed to be at least a millionaire. But the truth is different. Junior mining companies are usually underfinanced, always on the lookout for investors, often hyping it up. Mining investment, even in battery metals, is not considered attractive anymore. Instead, funds go to technology companies, no matter if they are in cleantech, biotech or fintech. Seed capital and venture capital companies in the vast majority choose to finance technology companies. Try to sell them the idea of more old-fashioned industries, such as shipping or mining, and they may not even be willing to evaluate your business plan—not because it does not make sense, but because it just does not fit their investment profile, which is skewed towards tech.

The smart money with an appetite for mining investment is concentrated only in a handful of places, mostly in Canada and Australia, where most globally active mining companies tend to be listed. So if your company has an office in London and a mining project in Congo, it would still probably make sense for you to list on the Australian Securities Exchange or the Toronto Stock Exchange. There are some small pockets of investors left in London, where the still-influential London Metals Exchange is based, but this is not where most mining companies' shares are listed. The London Metal Exchange is an institution through which financial contracts for deliveries of metals themselves are concluded. In fact, the London Metals Exchange, despite being based in Europe, belongs to Chinese capital. It was bought by the Hong Kong Stock Exchange, a happy development considering that, before Hong Kong took over, the risk of the London Metal Exchange becoming marginalized by its Asian counterparts was a lot higher.[106]

Maybe Western mining entrepreneurs were happy to sell their Latin American projects—which likely would not have been able to make it to the production phase if the Chinese had not bought them out—but the situation is concerning for some of the other stakeholders. So as not to be accused of being Western-centric, let us consider the Japanese, who are not very happy about the way things are. I once had a chance to talk with a Japanese executive, who said that he needed to look at European projects because in Latin America, Asia, Africa and Australia, Chinese money was everywhere—either in talks-in-progress or in done deals. My Japanese interlocutor did not want to enter the game, because he was looking for value offerings, not a bidding war with the Chinese that would end up in him overpaying, supposing he even won at all. There is another reason to bring up Japan as an example. Japanese companies are known to be sitting on an enormous pile of cash. As of 2019, according to filings, they are holding on to $4.8 trillion.[107] Just to give a comparison, and to illustrate how large this number is, Sweden's GDP for 2018 was $551 billion, roughly one eighth of the cash hoarded by Japan Inc. In addition, for the twenty-eighth year straight, Japan's net assets overseas were larger than those of any other nation. Net asset value is derived by subtracting the value of assets in Japan held by foreigners from the amount abroad owned by the Japanese government, companies and individuals. At the end of 2018, this net asset value was at $3.1 trillion.[108] Again, let's put this figure in context. Saudi Aramco, able to produce 12.5 million oil barrels per day[109]—by some analysts called the most profitable company on earth and the largest oil producer—briefly reached a valuation of $2 trillion on the stock exchange.[110] With this mighty spending power, the Japanese are, in theory, a formidable contender in the scramble for lithium assets. Yet still they did very poorly in comparison to the Chinese.

Japan's position on the sidelines is all the more surprising considering that it has a long history in the battery industry—in fact, Japan is the country that started it on a mass scale. Neighbouring South Korea, seeing Japan's first successes in the field, decided to develop a cluster of battery companies of its own, working closely with Japan to pick up the necessary know-how. From humble beginnings South Korea managed to build battery companies that dominate today's EV and consumer electronics supply chains, but, like Japan, it has stayed behind on lithium.

Nevertheless, one could argue that at least Japan and South Korea have built themselves an extremely strong position in the battery supply chain, even if they focus more on value added parts, such as the batteries themselves or components such as cathodes, anode materials and electrolytes. The US and European Union are in less enviable positions.

The US even had a head start, in terms of its contribution to the research and development of the battery. The most popular battery chemistry of today—NCM—was developed by the Argonne National Laboratory,[111] a lab with traditions going back to the Manhattan Project, where the best scientific minds raced against Nazi Germany to create the first atomic bomb. Surprisingly, NCM was developed and patented around 2000, so the technology has been there for some time, even if its newest reincarnation, as the 811 series, is considered to be the cutting edge of battery advancement. Yet, despite its advantage on the science front, the US has not managed to bring NCM cathode materials to production at anywhere near the same level as the Chinese, who nowadays have most of the market share.

Lithium's place within the battery is in the cathode materials. Other elements, such as cobalt, nickel, manganese and iron, are also present in different amounts, depending on the cathode chemistry. The unique status of lithium as a battery material comes from the fact that it is the only element that you will find in all available lithium-ion battery cathode materials.

While the US produces very little lithium, Canada produces none.[112] The United States Geological Service publishes authoritative mineral commodity summaries on a range of metals, with reliable data covering the whole world. Despite being a US publication, it keeps lithium production levels in the US a secret. This is because there is only one lithium producing mine in the US, and it belongs to a single American company: Albemarle. The agency refrains from publishing the data so as not to reveal Albemarle's sensitive commercial information. Albemarle is a lithium major—the largest lithium producing company in the world by volume.[113] The amounts of lithium that it produces, however, are not much higher than those of its main competitors: Ganfeng, Tianqi and, to a lesser extent, SQM. Albemarle is also not nearly as aggressive as the Chinese majors in securing assets abroad. The company seems to be content with their strong position in Australia and Chile. Albemarle's stated intention of creating the best value for its shareholders, as opposed to becoming the largest lithium producer, marks a clear distinction between the Chinese and American approaches.

Albemarle's raison d'être is to serve its investors. Tianqi and Ganfeng, like Albemarle, are public companies listed on the stock exchange, but they need to keep China's best interest in mind. They have remained to a large extent family companies, built from the bottom up on the bedrock of struggling state-owned facilities bought on the cheap. Naturally, they are run for profit, but profit is not their sole consideration. They must not forget where they came from—and how their expansion is supported. Thus, they must simultaneously keep both profits and national strategic considerations in mind, as large factors guiding their decisions. Ganfeng has stated that its ambition is to be the biggest lithium producer in the world—and has not accentuated serving shareholders as being its guiding principle as much as Albemarle. The lithium industry is yet another facet of state capitalism with Chinese characteristics—something that economists have been talking about for some time.

An interesting question is how the West and other large Asian economies are willing to compete with China. Will they stick to their established development models, or try to emulate the Chinese way—in whole or in part, or by coming up with some sort of synthesis of the different approaches? This is a question which not only concerns the battery supply chain, but is interesting to look at through the battery supply chain prism.

Trade war is one example of how the West is dealing with the problem of Chinese domination. Peter Navarro, who served as the White House trade adviser under the Trump administration, calls state subsidies one of China's 'seven deadly sins', and argues that these must be addressed before the two countries can normalize trade relations again.[114] The World Trade Organization (WTO), of which China went to great lengths to obtain membership, is also being used to exert pressure on the country. A group of US, EU and Japanese trade ministers within the WTO has met on numerous occasions to try to convince China to make state involvement in the economy more transparent, and hence one day to create a more level playing field for companies that do not receive that kind of support within their own countries. Notably, South Korea is not a part of this pressure group. The miracle of South Korea's economic growth was to a great extent the result of export-oriented industries in specific market segments that were designated and aided by its government.

The character of competition with China is multifaceted. One approach is to undermine the Chinese model through criticism. Another is to emulate it, within available legal frameworks. The latter is exactly what the EU and US have been doing. Senators Lisa Murkowski and Joe Manchin introduced bipartisan legislation addressing America's dependence on foreign countries for its mineral needs in the battery, cleantech, high tech and defence industries. 'Our nation's mineral security is a significant, urgent, and often ignored challenge. Our reliance on China and other nations for critical minerals costs us jobs, weakens our economic competitiveness, and leaves us at a geopolitical disadvantage,' said Murkowski.[115] The media has called Murkowski's act a comprehensive plan for fostering the domestic production of minerals considered critical to the United States.

Murkowski's initiative turns the discussion in the right direction and maps out an important problem, but it tackles issues that are not critical for the industry. Its key components of a nationwide resource assessment, permitting reforms, and a study of the nation's minerals workforce are hardly likely to change anything for the lithium juniors in the US.[116] In comparison with other countries, the US mining legal framework is extremely robust and business friendly. Yet mining projects still come to fruition in places where this is not the case, as long as there are funds available. How about resource assessment? Again, US mineral resources, in comparison with those of China or Argentina, are already better mapped out. Yet the US does not possess a similar number of projects to these countries. It is access to funds that North American lithium juniors really need.

For many battery metals mining entrepreneurs, Murkowski's legislation amounts to much ado about nothing. Arguably, the only real positive is that it brings good publicity and awareness to the mining industry—in a country where the new generation more readily associates 'mining' with data than minerals.

People often think of Tesla in the context of the battery supply chain. Tesla is an innovative company, doing for EVs what Apple has done for smartphones. Its products are considered cool, premium and very American. Tesla, contrary to common belief, does not make batteries. The battery, a key element for the functioning of electric cars, is not a Tesla product. For the Nevada-based gigafactory, it is in fact Japan's Panasonic that produces its cells. And the lithium-containing cathode material, a critical component of the cell and pretty much responsible for the battery's performance, is produced by another Japanese company, Sumitomo Metal Mining.

Cathode materials are the crystal structures where lithium compounds are inserted during the production process. During charging, lithium leaves the cathode, and when the cell finishes discharging it needs to be able to return to the crystal structure. The process repeats each time the cell is charged and discharged. On a nano level, the cathode material crystal structure needs to be robust enough to withstand these lithium departures and comebacks.

It is important to understand the difference between a cell and a battery, and to use these terms correctly. Tesla vehicles' source of power, a battery pack, is made of batteries, and batteries are single cells connected together in modules. The cylindrical cells that you will find in a Tesla product would, on first impression, look very similar to the cells you buy to power a TV remote. They just have superior electrochemical properties and come in large numbers within the battery pack—say, around 7,000 cells connected into cell modules, together with the system that manages them, creating a battery pack.

The European Union used to have a curious stance on batteries. Its decision makers thought that batteries were going to soon become purely a commodity business, and hence not a valid point of interest for the advanced, value-added-oriented economies of the old continent. The EU was also worried about the creeping overcapacity build-up—perhaps a valid anxiety, and one that we will return to later. Still, the EU missed a few important counter-arguments that might have tipped the scale towards building battery factories at home. Firstly, automakers favour just-in-time deliveries, and are used to working with clusters of industry placed nearby, preferring to have a supply chain for key components localized close to the centres of demand. To date Germany, Switzerland, Hungary and the Czech Republic have built a massive number of successful, specialized companies serving German auto-industry giants such as BMW, Mercedes, Audi or Volkswagen.

Secondly, even if the EU initially did not want battery factories, it still wanted to lead with innovations in the field. It is widely accepted, however, that to make the most of the natural synergy between industry and research, it is easier by far to have a manufacturing base in geographic proximity to the labs. It also gives universities spinning out start-ups based on research in academia the chance to recruit clients from the industrial players in the region.

Take Germany's state of Bavaria, for example. Its sprawling start-up scene is built around the automotive industry that operates in the geographical vicinity, and the R&D activities in locally-based academic institutions, such as Munich's Technical University or the Max Planck Institute, are carried out in close cooperation with local manufacturing companies and start-ups. There is no reason why this success story could not be replicated somewhere else with the battery industry.

No one thought like that in 2015, the year when Li-Tec, a battery producer owned by Daimler (the company behind the Mercedes brand) had to close down.[117] Li-Tec had an impressive technology for that time, but low volumes made its production costs expensive, and this sealed its fate. In early 2016, Dieter Zetsche, Daimler's CEO, decided not to proceed in a joint venture with BMW and Audi to invest in battery cell production in Germany. He told the press that 'there is de facto a massive overcapacity in the market today and cells have become a commodity. The dumbest thing we could do is to add to that overcapacity'.[118]

Luckily, the EU has since started slowly to wake up to the new reality. In October 2017, at an EU gathering, the vice president of the European Commission said: 'The lack of a domestic, European cell manufacturing base jeopardises the position of EU industrial customers because of the security of the supply chain, increased costs due to transportation, time delays, weaker quality control or limitations on the design'.[119] Executives from BASF, Daimler, Renault and Umicore attended the meeting. Egbert Lox, vice president of Umicore, said that Europe needed an 'Airbus for batteries'—a somewhat self-serving observation, considering the company's strong position as a cathode material supplier.[120]

It is, in fact, the South Koreans who have positioned themselves, above Europe, to profit from the nascent European EV industry. In 2015, LG Chem took the decision to build a battery plant in Europe, targeting a site near Wroclaw, Poland. The location was not surprising considering its proximity to the German automotive industry, the lower production costs compared with those in Germany, and LG's long history in Poland. In fact, LG's exposure to the Polish market reaches back to deals with the controversial Art-B in the wild days of Poland's postcommunist transformation. The story goes that the Art-B founders, still in their twenties, made a sensation in South Korean business circles when they paid Hyundai for 50,000 cars filling two container vessels, by simply handing over a credit card during the meeting.[121]

The plant has been operational since 2018,[122] and Swedish Northvolt, the only truly European EV battery manufacturer, is scheduled to start commercial operations in 2021.[123] Another South Korean company—LG Chem's main domestic rival, Samsung SDI—established an EV battery plant in Hungary, also operational from 2018.[124] But the story of Samsung SDI in Hungary hit a setback—and an interesting twist—when EU antitrust regulators started investigating the Hungarian government's plan to grant 108 million euros to Samsung SDI's battery production facility.[125] According to Brussels, this may have constituted unlawful state aid, defined as financial assistance given by the government to companies that have the potential to distort market competition within the EU's single market. Under EU rules, state aid is generally not allowed. Member-state governments can provide it only with approval of the European Commission. So the EU, on the one hand, wants the battery sector to develop, but on the other hand launches an investigation as soon as one of its members puts forward a concrete incentive for the top producer to bring its know-how and demand for the battery components—not to mention jobs and tax revenues—into the region.

Swedish Northvolt has proven so far to be a project with huge potential. Its co-founder Peter Carlsson's success derives not so much from the idea to build a battery factory in Europe, but from its execution. He managed to obtain massive funding for a massive project, selling it on a wave of climate change concerns, and has already managed to secure the buyers for its future production.

Mr Carlsson has been immersed deep in high-tech industry supply chains for some time. He is equipped with the necessary skills and a good network, having worked as a head of sourcing for Sony Ericsson, head of purchasing for NXP Semiconductors in Singapore, and VP of the supply chain for Tesla.[126] He raised capital for his battery plant from large players such as Volkswagen, Goldman Sachs, Swedish pension funds, the IKEA-linked IMAS Foundation, and the European Investment Bank with a €350 million loan.[127] Building battery plants does not come cheap. LG Chem's plant in Wroclaw will have cost €2.8 billion by the time it reaches its planned capacity of 70 GWh per year.[128] Northvolt's capacity is planned to reach 32 GWh.[129] One GWh is a unit of energy representing one billion (1,000,000,000) watt hours. Tesla's popular Model 3 has a 40 KWh (kilowatt-hour) battery as a standard offering. Northvolt's 32 GWh capacity per annum means that if the factory maximizes its production yields, it should be able to output cells for around 640,000 Model 3s annually. Bearing in mind that Tesla sold only 367,500 cars in total in 2019, this brings us to the question of overcapacity.[130] If one looks at all future battery-production capacities announced globally, we enter this decade with already around 2 TWh (terawatt-hour) of capacity in the pipeline.[131] For the sake of this exercise, and again taking Tesla's Model 3 as an example, this is enough to equip 40 million EVs with batteries, annually. Production of all passenger and commercial vehicles in 2018 (both conventionally powered and electric) totalled around 97 million units.[132] But in 2019, global EV sales were just at around 2 million units.[133] Does this mean that the market will face a massive overcapacity problem? Is there any sense in building yet another factory?

To answer this question, first we need to understand that these are just announcements and predictions. For many of the industry's newcomers, the planned factories may never materialize. If we look at announcements by the main battery producers, which you can easily count on the fingers of two hands (LG Chem, Samsung SDI, BYD, CATL, SK Innovation, Panasonic and maybe a few others), the capacity in the pipeline is much, much less impressive. And established EV manufacturers will not risk contracting untested battery suppliers, due to safety and performance risks that could potentially deal a fatal blow to their brand. Just imagine any famous car brand's vehicle catching fire in a minor traffic accident, incinerating the driver as the result of a battery explosion. Moreover, EV batteries are always developed in close cooperation with an EV manufacturer, since the technology needs to be adjusted to a given EV model. Specifications for the batteries and components needed are always a closely guarded secret. Therefore, the situation is not parallel to a commodities market, where an industrial buyer announces a tender for required standardized goods and whoever bids the lowest price and best payment terms wins. Contrary to what Daimler's CEO said, batteries have not been 'commoditized'.[134] It is also unlikely that they will commoditize anytime soon.

Having discussed the European battery sector, let us now look at the European lithium mining industry, still in its infancy. European Metals, an Australian company exploring Czech mineral deposits, is an interesting example of the horse trading that goes on between a private player with the know-how but not all that much money, and a European government with money but no idea of how to proceed with a resource that happens to be found within its territory. European Metals bought exploration rights around Cinovec in the Czech Republic in 2012, and planned to start mining by 2019, assuming a nameplate capacity of around 20,000 tons of lithium carbonate equivalent.[135] The company forecasted that it would have to spend around $393 million to bring the mine online.[136] Much like battery manufacturing, lithium mining is highly capital-intensive, but the eventual rewards for taking the risk can be large. A pre-feasibility study for the Cinovec project determined the cost of production at $3,483 per ton.[137] In today's depressed market, a ton of battery-grade carbonate can easily achieve $8,000. During a good market, prices could rise to much more than $10,000 per ton. This implies a gross margin of between $4,500 and $8,000 per ton. In the summer of 2017, everything looked bright for Keith Coughlan, the managing director of European Metals who boasted thirty years of experience in the finance and resource industries.[138] 'The permitting process is going along as it should be, everything is happening in an orderly fashion,' he said in an interview with Reuters.[139] There were just a few permits left to obtain from the Czech mining and environmental ministry. The presidential campaign was beginning in the autumn, but political risk was not something that the company could reasonably expect to impact its business. The Czech Republic was by then a stable, mature democracy and a member of the European Union.

But in 2017, a wave of populism was sweeping through Europe. A year earlier, the UK had voted to leave the EU, and in Poland, the Czech Republic's neighbour, thousands had protested against the government's attempts to secure full control over the judiciary. In the autumn of 2017, Andrej Babiš, the charismatic leader of the populist ANO party, won the Czech parliamentary election by a landslide. Babiš is a man of many talents. He had made a fortune starting out in fertilizers and ending up with a large conglomerate that had interests in everything from chemicals to the mass media industry, making him the second wealthiest man in the Czech Republic.[140] Allegedly he also dabbled in working as an agent for StB, the Czech intelligence and counter-intelligence agency during the communist era (a claim that he vigorously denies).

The commercially astute Babiš could likely not fathom why in the world his country should be ceding the rights to a key energy resource of the future to an Australian start-up. He decided to take measures. After a charismatic speech in parliament, in which he accused his predecessors of selling out Czech national interests,[141] he managed to secure the support of the coalition, and the Czech executive branch was instructed to treat arrangements with the Australian lithium company as void.[142] European Metals still had legal titles to the resource, so recourse to a supranational court was possible. But most likely the company did not want an uphill struggle on what quickly came to be seen as an 'enemy territory'. In the meantime, the lithium market turned sour and financing became tighter. Investors were not all that willing to support the company while it was in conflict with the government—as such a conflict rarely ends well. Finally, European Metals gave in to the pressure and reached an agreement with CEZ, the Czech state-owned power company. Under the conditional agreement, European Metals received a loan, with CEZ reserving rights to convert the loan into shares, thus obtaining a controlling stake in the company.[143] On the surface it looked like a win-win—the company raised cash when finances were tough, and the Czech government secured the right of control and a share in the company's earnings if the project turns out to be successful. The Czech lithium saga continues. The mine is still in its very early stages, while CEZ is making far-fetched plans to enter the energy storage market with domestically sourced lithium.[144] The Czech example is a good demonstration of how it is not only the world's great powers—China, the US or the EU bloc—that have acted to secure their interests in the new energy sector.

Is it enough, though, for Europe to mine lithium and construct battery factories to create an independent, cost-efficient battery supply chain, or is there a missing piece to the puzzle?

The problem with lithium is that it needs to be processed in order to be used in a battery. China, as described at the beginning of this chapter, converts raw material sourced in Australia into the compounds used in a battery. Unlike Australia, however, Europe is at a considerable distance from China. It takes a vessel around ten days to travel between Australia and China, versus around thirty days to travel between China and Europe. Around 60 per cent of marine vessel operating costs relate to the consumption of the bunker fuel that propels the ship.[145] Large, fully electric vessels are not yet even on the engineers' drawing boards. Bunker is an expensive oil product, and when it is burned, CO_2 emissions are created. Thus it is time consuming, costly, and polluting to transport European-mined lithium to Asia to be processed and then to shuttle it back to European battery factories—definitely not an ideal solution. To add the missing pieces of the new energy puzzle and secure the localized supply chains that automakers and politicians are aiming for, Europe and the US need to build conversion and battery component facilities. This is not an easy feat, as the production of lithium chemical from ore can be challenging for even the most experienced lithium companies. It took Tianqi Lithium over three years and $400 million to construct a processing plant in Kwinana, Australia, and make it partly operational.[146] It is likely to take another twelve to eighteen months of effort to bring it to a first phase annual capacity of 24,000 tons of lithium hydroxide.[147]

Tesla's Model S, used in our example above, consumes roughly 55kg of lithium hydroxide per vehicle, meaning that Tianqi Kwinana's capacity would be enough to equip around 430,000 units with batteries annually.[148] This represents a large part of the current market share, considering that around 2 million EVs were sold in the world in 2019.[149] Based on the fact that nameplate capacities are almost never used to their full extent, and on the assumption that the world's automotive industry goes only one way—electric—Tianqi does not stop there. In the future the company plans to bring its capacity up to 48,000 tons of lithium hydroxide per annum, at an additional cost of $205 million.[150]

Some European lithium producers do plan to set up conversion facilities and produce hydroxide adjacent to their mines. But considering how much money and time it took Tianqi to do that, I do not think their plans will materialize quickly. Nevertheless, Europe needs badly to have at least one such facility if it wishes to get further along the path to supply chain independence. The old continent has already overtaken the US in the number of EV sales, by over 270,000 units in 2019 alone.[151]

Having discussed the missing puzzle piece of converting raw lithium ore, we can move on to the next one—namely, the production of cathode materials. This puzzle piece is only partially missing in Europe, thanks to two companies—German BASF and Belgian Umicore. Umicore, through its factories in South Korea and China, and BASF through its joint venture with Japanese Toda and factories in the US and Japan, fairly dominate the global cathode materials industry. The majority of their customers are battery makers from South Korea, China and Japan.

BASF did not invent NCM, its best-selling cathode material for EVs. The US-based Argonne National Laboratory formulated this revolutionary, layered crystal structure made with different ratios of nickel, cobalt, manganese and lithium. Back in the day, they were worried about protecting their invention from being copied with no regard to patents by the sprawling Asian battery industry.[152] So Argonne came up with an ingenious idea, to create an artificial scarcity on the market by selling licenses to large players who had the ability to successfully resell them, as well as the financial means to defend the patents abroad. This strategy led to two large corporations—BASF and Toda— becoming exclusive worldwide licensees.[153]

BASF joined forces with Japanese Toda in 2015 in order to increase its manufacturing capabilities, as Toda had great practical experience in working with crystal materials at nanoscale, as the result of its years of operation in the iron oxide industry. Partnership with Toda also guaranteed access to the growing Japanese battery market.

Meanwhile Umicore started manufacturing and selling NCM cathode materials to the battery industry, believing that their formulation of NCM was unique and did not require a license from BASF. After a short legal dispute, the International Trade Commission determined that Umicore was infringing the patents, and whoever used the cathode material from Umicore in its batteries was also committing an infringement.[154] Umicore did not have much choice other than to buy a license and thus settle the legal case with BASF and Argonne.[155]

By now, the two companies have put the disputes behind them, and, with experience and manufacturing processes developed in Asia, have positioned themselves to take the European cathode market by storm. Umicore is planning to open its NCM cathode production site in Nysa, Poland, in 2021, which is well placed to supply the Wroclaw LG Chem factory.[156] Meanwhile BASF revealed its plans to build a production site close to the planned European Tesla gigafactory near Berlin.[157]

For some countries, a lack of lithium resources or any sizable EV market to speak of is not an impediment to entering the lithium game. With an eye to the future, India is counted among the emerging global powers, due to the size of its military and economy. It is already the fifth largest economy by nominal GDP size, placed in rank between Germany and the UK. But with its population of over 1.3 billion, it is poised to do even better.

In India, a little over 1,500 EVs were sold in 2019.[158] No lithium deposits have ever been discovered within the country's borders. But despite this, it is crucial for India to one day secure cheap access to electric mobility for its growing population, particularly considering the need to curb the current pollution levels and the country's dependency on oil imports. According to Greenpeace, twenty-two of the world's thirty most polluted cities are in India, including Delhi, its capital.[159] India also consistently ranks as the world's third largest importer of oil, behind China and the United States, with foreign oil responsible for around 85 per cent of the country's needs.[160] And much of India's oil imports come from politically volatile jurisdictions—in 2018, Iraq, Iran and Venezuela ranked among India's top five crude oil suppliers. In 2017, India's prime minister, Narendra Modi, proposed that by 2030 the sale of new conventional internal combustion fuel vehicles would be banned. But this is in line with small countries such as Denmark and Iceland, who already lead EV sales in Europe—so this over-enthusiastic target has since been revised to a more reasonable goal, of one third of new vehicles sold in India being electric by 2030. A range of subsidies programmes and tax breaks has been rolled out to support this transition. On a political level, India has started to look in the direction of South America, a part of the globe where Indian interests have historically been very limited. During 2019, India's high-level politicians met with representatives of all three countries of a 'Lithium Triangle': Argentina, Bolivia and Chile, with the Bolivian visit marking the first Indian government visit to this country in history.[161]

The Indian government has directed three state-owned mineral companies (National Aluminium Company, Hindustan Copper and Mineral Exploration Corp.) to join efforts and create a special vehicle for the acquisition of lithium and cobalt resources abroad.[162] In the eyes of India's politicians, the major hurdle to overcome for an electric revolution in India is not lack of battery factories, as those can be licensed and built even by domestic companies. The local automotive industry, with strong brands such as TATA, potentially has the capability to one day produce affordable electric cars for the Indian middle class. But the precious resource of lithium itself cannot be created out of thin air—it is either there in the ground or it is not. This is why India has turned its gaze to the South American Lithium Triangle, which holds lithium resources of 47 million tons out of 80 million tons globally.[163] We will look at the Lithium Triangle in the next chapter, to find out how a large amount of new oil in the ground, a difficult recent history, foreign influences, and local coteries all play a role in determining the future of the region.

3

LITHIUM TRIANGLE

The salt flat of Atacama is one of the least habitable places on the planet. It is arid, with an average annual rainfall of below 2mm per year. Even the Sahara can experience up to 100mm of rainfall in its 'rainier' years. London, where this book is being written as raindrops hit the windows to the tune of 'Sketches of Spain', experiences 583mm of rain spread over 106 days of the year on average.

The lithium resource contained within Salar de Atacama—to use its proper name—is closely guarded by the Andes mountains, with the Cordillera Domeyko range, named after my Polish compatriot who did his fair share of mapping Chile's mineral resources, closing in from the west.

The place is rugged, in the true sense of the word, and desolate. But there is an otherworldly beauty to it, with terrain and skies bursting with hues right before sunset, making you feel as if you are on the surface of another planet. This unfriendly landscape is home to a bird with equally atypical colours—the pink flamingo—which has learned how to prosper from the salar. The bird skilfully extracts the algae and brine shrimp full of

85

carotenoid pigments, to which it owes its colour, from under-neath the surface.

The protagonist of this chapter is as much of a rare bird as a pink flamingo. A controversial figure, an éminence grise, a man of influence who, despite being the former son-in-law of Chile's late dictator, never saw himself as a part of the country's estab-lishment. Mired in scandals and allegations involving everything from stock manipulations[1] to the creation of complex offshore structures bringing to mind those featured in the Panama Papers,[2] to influencing the country's political landscape by financing electoral campaigns, he is also father to Chile's success on the lithium market.[3] Julio Ponce Lerou, with his elongated figure and a face lively with expression, suits the smooth opera-tor role that he has been playing for years in Chile's business and political circles. His influence has been felt throughout Chile's lithium sector for decades, and the story of his life is closely intertwined with the country's recent political and eco-nomic history.

Julio was born into a solidly middle class family of medics. It was thanks to the proximity of his family's property to that of General Augusto Pinochet that he met his future wife.[4] In his own words, the links to and friendship with his father-in-law, lasting even beyond the failure of the marriage itself after many years, was the most 'expensive' relationship of his life,[5] costing him dearly in terms of the mental toll it took.[6] Julio never saw his success as a product of this relationship, despite the many positions he held in state-related companies and agencies.[7] And who could ever ascertain the extent to which this belief was simply a mirage, driven by the wish of a strong character to join the ranks of 'self-mades', or whether it was in any way grounded in reality?

Wood and wood pulp have long been, and continue to be, one of Chile's top exports, along with metals, fish, fruit and chemicals. So it is no wonder that ambitious young Julio graduated as a forest engineer. He made his first commercial success developing a forestry business in Panama and earned his first sizable amounts of hard currency there, being compensated on a commission basis in addition to a fixed salary.[8] He was in Panama in the days when General Pinochet made his grab for power, in the era before Twitter, cut off from much of what was really happening on the streets of Santiago. Ponce was worried for Pinochet's wellbeing in the coup, given his exposure as a top general, even though he was completely unaware that in fact it was his father-in-law who was leading it.[9]

A picture of the historical background will help readers to fully understand the current socio-economic situation in Chile and its lithium sector. Chile's recent history and national psyche have been scarred by years of dictatorship, when draconian measures were exerted on any form of dissent. Even though Pinochet's rule ended in 1990, the process of reckoning with those troubled years and the structures that the regime left behind are far from finished.

In the early 70s, Chile was considered a model of stability in Latin America, before social discontent provided fertile ground for a regime change. To shed light on the 1973 coup d'état, we need to introduce three interested parties: the rightfully elected president, Salvador Allende, the Chilean Congress, and the US government under President Richard Nixon's administration. Allende was extremely left wing, being in fact the first Marxist ever elected in a real democracy. One of the founding fathers of the Chile Socialist Party, he tried a number of times to get elected, running for president in 1952, 1958, 1964, and finally winning in the 1970 election by a narrow margin. Back then, if the presidential election was won by a small percentage, Congress had to decide on who became president. At the peak of the Cold War, it was not advisable to be an openly Marxist candidate in a region the US considered its own backyard. In the election of 1970, the CIA had its most favoured and least favoured candidates among the three that were standing for Congress's final vote. Needless to say, Allende was the latter. The incumbent president, Montalva, was the Agency's go-to man.[10] Despite not being the exact embodiment of a free market advocate, with his progressive tax reforms and plans to nationalize Chile's copper industry, he was seen by the US as a safe choice at a time when Latin America was plagued by military juntas. Since Chile's constitution forbade the president from being elected for a second term in a row, the idea was for the third candidate, Allessandri, to be elected, only to resign shortly after and make way for a new election during which Montalva would be expected to win legitimately. This elaborate conspiracy did not work, though, since Allessandri divulged his plans of early resignation to the congressmen. Outraged, Congress reluctantly voted for Allende.

During three years in power, Allende started to make moves which, in the opinion of foreign policy analysts, might have ended with Chile turning into a communist country. The CIA's worst fears started to take shape. Chile began to nationalize its main companies on a large scale, turned healthcare and educational systems public, started to redistribute land by taking it away from large landowners, and finally invited Fidel Castro for a high profile visit—all under the outspoken banner of the 'Chilean Way to Socialism' programme. But the one thing the Agency had feared—an economically successful model of socialism in Latin America, one that would encourage similar scenarios in neighbouring states—did not materialize. Instead, inflation went berserk, reaching 140 per cent annually. People, especially small business owners, went on the street. A precariously unstable socio-economic environment emerged, opening the way for Pinochet's coup.

Fidel Castro gave a piece of advice to Allende: keep the military close. Unfortunately for Allende, he failed to appreciate the value of this insight, perhaps until it was too late. But in any case, the Chilean officers had created a caste all of their own. They were mostly apolitical; and if political at all, then broadly right wing. They partied together and they went on vacations together, closing themselves off in army-owned holiday resorts. Army families even tended to marry each other—the marriage of Julio Ponce and Veronica Pinochet was an exception to the rule. But what is perhaps most important in this story is that the military personnel had been much underfunded. They looked with envy at the lavish lifestyle of their counterparts in neighbouring countries, where military juntas had taken over. Such moods were dangerous; they just needed fertile ground and a spark for a coup d'état to take place.

This spark came from the growing division between the executive and legislative bodies, government and the congress. The government under Allende wished to use the socio-economic turmoil as an excuse to consolidate its power for the purpose of bringing back social order. Congress made a call for the military to oppose that. On 11 September 1973, major army dignitaries, including Pinochet, stopped answering phone calls from the anxious president and his minister of defence. Soon, the country was under military control and its radio and TV stations were shut down.

Allende and his closest bodyguards had been staying in Chile's humble presidential palace, La Moneda. Despite the dire economic situation in the country, he still commanded strong support. Chilean society was divided, like never before and never since. The last thing Allende wanted was a violent struggle, a civil war. He refused to flee or to call upon his supporters, staying in the palace even after being informed of its imminent bombardment. Until the end he refused to back down, relying on the legitimacy afforded him by his democratically won election. In the end he chose a bullet, fired from an AK-47 he had reportedly received as a gift from Castro. Pinochet's rule, meanwhile, was full of concentration camps on the same Atacama Desert where the country's lithium is to be found; mass executions were frequent, as were the disappearances of left-wing political opponents.

Julio Ponce is a natural born entrepreneur. Already as a teenager, before he started a relationship with Maria Veronica Pinochet, he procured fish from a local fisherman and sold it to his future mother-in-law.[11] The Valparaíso region where he grew up, and where his future father-in-law was stationed with his family, is a beautiful coastal region of Chile, full of spas and pristine landscapes. At the same time, however, it must also have been a little bit on the rough side. Julio Ponce once remarked that, being a product of the public education system, he is afraid of neither poverty nor jail.[12] This down-to-earth, middle-class background, together with his effectiveness in professional affairs, was the ground on which Ponce built an excellent rapport with Pinochet, who, unlike many among Chile's ruling classes who hailed from aristocratic, industrialist or land owning families, came from similar social circles.[13]

Julio Ponce has given only three full interviews to the press. Sometimes, being publicity shy provides a person of influence with an aura of mystery that provokes gossip and speculation that feeds on itself. In private, Julio Ponce apparently likes to quote Oscar Wilde: 'There is only one thing in life worse than being talked about, and that is not being talked about'—is reportedly one of his favourite sayings.[14]

Despite having a close relationship with his father-in-law, Ponce hated to be associated with him. The public and media referred to him just as *el yerno*—the son-in-law. Being a very sociable, seemingly laid back character, he disliked how the attitude of the people around him changed when they discovered his relationship to the dictator.[15] He liked to joke a lot and despised all sorts of titles and formalities. When he came to work for

Conaf (Chile's National Forest Corporation)—which, through a network of state-owned companies, controlled much of the country's forest and wood industry—he corrected employees who referred to him as *el director* and instructed them to call him by his name.[16] At the same time, however, he remained fiercely loyal to his wife's family. In the late 80s Pinochet's popularity was in free fall. When somebody around the table at SQM made a distasteful joke about the dictator and his wife, Julio Ponce retorted that many had died for much less than this.[17]

In total, Ponce spent over 3,300 days working for Pinochet's government, heading up Conaf and Corfo (Production Development Corporation), another important Chilean government agency that even today holds sway over the Chilean lithium sector.[18] It was his employment in government that made him who he is.

Ponce was also successful in his posts in private companies within the forest industry. He held executive positions in Chile and Panama when he was only in his twenties. This situation is hard to imagine in today's corporate environment, yet was completely normal for graduates of good universities at that time in Latin America. But in terms of building a network and experience across a range of industries, nothing could match governmental posts. For Ponce, the decision to go into the public sector was not about the money. By accepting a role in the Chilean public sector he earned 90 per cent less than at his job in Panama.[19] But, young and idealistic, he might have believed that with experience accumulated while working for private players abroad, he would be able to reform the inefficient, national behemoth of Conaf, which had run up extremely high costs during the period of socialist government. He saw himself as a man with a mission, in charge of the country's strategic industry from an early age.[20]

After taking the job, he began by reducing staff numbers—not only laying off employees in redundant roles and closing inefficient facilities, but also removing all of those in managerial positions that opposed his vision, thus consolidating his power.[21] Ponce was instrumental in opening Chile's forestry sector to the world, exporting Chilean wood to a number of new markets and attracting foreign investments.

Heading Conaf and later Corfo gave Ponce enormous influence. Through these agencies, he held fifteen directorial-level positions in state-owned companies, in areas as diverse as the sugar, chemicals, oil, telecommunications, forestry, mineral and electricity industries.[22] He later had to explain himself in the media, clarifying that those positions were never financially compensated; he received only one salary at the agency, and these numerous posts were held across many years, not all simultaneously.[23]

Before we delve any further, we perhaps need to understand how unique Corfo is as an institution. You will not find an organization with a similar character in any other emerging market. It was founded in 1939 to change the economic landscape in Chile. Most of the largest companies in the country, in strategic industries such as oil and gas, chemicals or telecommunication, were initiated by Corfo, and the agency kept a stake in them for many years before the privatization programme started to take off at the end of the Pinochet era. Corfo played an enormous role in industrializing Chile, taking it from a poor country to a member of the OECD. Even now, it has an important part in pushing the economy forward by operating one of the largest venture capital funds in the emerging markets, investing in technological companies and providing massive amounts of loan guarantees to small and medium size businesses to boost their growth.

Ponce headed Conaf from 1974—the year after Pinochet took power—and Corfo from 1979, until 1983, when he had to step down amid allegations of corruption.[24] Shortly after, Ponce went into the agricultural and forestry business. He used his network at Corfo and operational knowledge of the agency to arrange a loan for his new company to acquire livestock.[25] That loan was only paid up in small part, after his business went bust.[26]

Ponce, meanwhile, paid for the business failure with his health, and suffered a severe mental breakdown.[27] The central thread of our story, however, is his period as a director of SQM. He held this position in a long stretch lasting from 1987, through a regime change and SQM's entry into the lithium sector, to 2015, when he had to step down under scrutiny over yet another affair.[28]

Ponce not only managed to lead SQM, one of the lithium sector's largest producers, through many years, but also became the company's largest shareholder after its privatization.[29] Chile underwent a wave of privatizations in the late 80s. The idea was to turn state-owned companies into private companies, with the majority of shares owned by a minority of shareholders—crucially, the company's own labourers and the employees' pension funds.

What happened later resembled the process that took place in Russia and, to a lesser extent, in Eastern Europe. Labourers with shares were either incentivized or coerced to sell them, often below their fair value, to individuals or organizations who were really taking over their ownership. In Russia, instances of labourers giving up the vouchers exchangeable for shares in major Soviet industrial enterprises in exchange for bottles of vodka, or for any amount of hard cash they could lay their hands on amidst raging inflation, were disturbingly common. Chilean society, much more exposed to the inner workings of capitalism, was less naive regarding the shares' true value, but the mechanism through which a select few acquired vast wealth thanks to privatization ended up being fairly similar.

Ponce's control of SQM was secured through a network of offshore structures.[30] These are a waterfall of companies registered in tax havens, where one company owns the other so as to obscure their true ownership, and to allow for a greater exercise of control with a lower application of funds, by means of loans and external investors' contributions. The end entities, which figure as owners for these chains of companies, are often trusts.[31]

Trust is an interesting concept that goes back to medieval England, when knights leaving for crusades required a legal structure that would allow a nominated manager (trustee) to govern their wealth and property to the benefit of their families while they were away, or, indeed, dead. Nowadays, well-off individuals use these legal structures by transferring the titles of ownership to different assets such as stocks, real estate and yachts to a trust. The trust is managed by lawyers who are obliged to govern it strictly in accordance with guidelines specified in the documents that established the trust, and to the benefit of whomever the trust appoints as beneficiary. The beneficiary can be the person who transferred the wealth to the trust in the first place—but the wealth is no longer his in the eyes of the law. It belongs to the trust and he is merely the beneficiary. Thus, in case of a divorce or litigation, the beneficiary's assets stay safe and out of the reach of law. Julio Ponce's ownership in SQM is rooted in the Pacific Trust, registered in the British Virgin Islands.[32] From Pacific Trust stems a waterfall of offshore entities leading to the ownership of SQM shares.

It is a truly Byzantine structure, spanning different ty of entities and legal jurisdictions.[33] It would have evaded public scrutiny, had it not been for an investigation by Chile's financial regulator (SVS).[34] In 2013, Ponce was accused of forbidden stock manipulation practices[35] that led to losses for minority retail investors and Chilean pension fund shareholders.[36] Some of the entities in this structure, such as Oro Blanco or Nitratos de Chile, were regular companies listed on the Chilean stock exchange in which everybody could invest.[37] Transactions between those listed companies started to take place for which there was no business justification, happening within such short time intervals that there was no room for disoriented minority shareholders to react.[38] Namely, the shares were sold on the stock market and then bought back at a higher price—to the benefit of the company that bought low and sold high.[39] The harm obviously comes from the fact that if you invested as a minority shareholder in the company that was buying back the shares at a higher price, you were participating in the losses.[40] Every director of a stock-market listed company should act to benefit the company and hence its shareholders; it is their statutory duty. The decision makers behind the scheme acted to enrich themselves, instead of seeking to enrich the company's shareholders. Ponce received a fine to the tune of $70 million for enriching himself at the expense of other shareholders in a series of holding companies that controlled SQM.[41]

The waterfall case also managed to shed light on Ponce's links to Chile's political scene. Sebastián Piñera was Chile's president in 2010–14, before losing to Michelle Bachelet, only to return to the presidential palace in 2018, with his new term scheduled to run until 2022. His power derives not only from politics, but also from his personal and family wealth. Piñera's father was Chile's ambassador to the UN, while the young Sebastián grew up in Belgium and New York. He owns a natural reserve on the south end of Chiloé Island,[42] among other assets that amount to $2.8 billion[43]—a mere $300 million below the estimate for Donald Trump's personal wealth in March 2020.[44] He personifies the patrician pedigree so prevalent in Chile's circles of power, and that Julio Ponce never identified with. Educated at Harvard, he was the first right-wing president to win an election after Pinochet. Together with Bachelet he has managed to dominate Chile's political scene for decades.

In 1998, Piñera opposed the arrest and detention of Augusto Pinochet in London.[45] This does not mean that he and Ponce are friends, though. At least not anymore. During the SVS

(Chile's financial regulator) investigation, Ponce said on the record: 'If his Excellency the President had not participated in the Waterfall Case, there would be no Waterfall Case.'[46] The statement went viral in the media. In fact, it was not the first time Piñera had been accused of stock machinations. In July 2007, he was fined around $700,000 by SVS for purchasing LAN Airlines stock in mid-2006, a company where he was a director, while in possession of privileged information.[47]

Meanwhile, SQM has also been fined for illegally channelling funds to political parties in the years 2008–15,[48] for recording payments to entities associated with politicians for consulting, and for professional services that were never received.[49] The lithium producer has been known to bankroll Chile's political scene, both the left and the right.[50] Most of those donations have been entirely legitimate.[51] But nevertheless, these links to politics and the atypical ownership structure make SQM an unusual animal in the lithium zoo.

Lithium is not likely to become Chile's most important metal any time soon. Copper exports tend to represent around 50 per cent of all of Chile's exports by value, with some variance depending on how well the world's copper prices are doing in a given year.[52] Chile's lithium exports in 2018 amounted to $949 million,[53] roughly 1.25 per cent of the total export value. Nevertheless, it is the lithium that makes the headlines and sparks imaginations. At today's market prices, 9 million tonnes of Chile's lithium resources is worth around $526 billion.[54] The total value of Chile's exports was around $75.5 billion in 2018.[55] Saudi Arabia's exports were worth $294.5 billion, around 80 per cent of which was oil.[56] Large numbers start to talk only if we put them in relation to each other. When Tim McCutcheon, CEO of Wealth Minerals, said: 'Chile is essentially the "Saudi Arabia of Lithium"', he may have got investors all excited.[57] Numbers often take the excitement away, but usually they are better at describing economic reality than catchy phrases. Even if we took all the lithium that Chile has in the ground and sold it, we would still earn less than Saudi Arabia's three years' worth of oil exports. Chile may well be a Saudi Arabia of lithium, but Chile will never be a Saudi Arabia. Analogically, lithium has the potential to make Chile richer, but lithium alone will not make Chile rich.

Salar de Atacama, with which we started this chapter, is the world's premiere lithium resource. The lithium concentration expressed as percentage by weight (wt%) at 0.15 is the highest among the world's brine resources.[58] It means that in every 1kg of brine, there is 1.5g of lithium. Almost everywhere else in the world, the brine resources are of lower quality: Bolivia's brines measure at 0.045 wt% and China's brines at 0.03–0.1 wt% on average.[59] Salar de Atacama is also one of the highest situated salt pans in terms of altitude. Its extremely low precipitation and high exposure to the sun allow for smooth evaporation, a key step in lithium extraction from salts. These excellent environmental conditions for lithium extraction, a random gift of nature to Chile, allow for year-round, low-cost production, with a cash cost per tonne of finished product around the $2500 level.[60]

So what does all this say about the contribution that Ponce and his entourage made to SQM's success on the lithium market? If one's country is endowed with one of the greatest lithium resources in the world, what can possibly go wrong? Yet I would refrain from judging that Ponce had this success handed to him on a plate. Bolivia undoubtedly has the largest lithium resource in the world; nevertheless, it wasted many years on its efforts to commercialize it. As I argue in the next chapter, it was not for lack of funds, but rather a lack of talent and managerial skills, that Bolivia's efforts failed. Even the harshest of Ponce's domestic critics acknowledge that he deserves the credit for going into a business that almost nobody foresaw expanding on the scale that it did—and for making it work.

Lithium is still just one of the business lines of SQM, but it is the one that is chiefly driving the interest of global investors, who link expectations of the company's financial performance to the situation on the lithium market. In 2018, the lithium business arm was responsible for 53 per cent of the company's gross profit; thus, this approach has a strong basis in the numbers.[61]

SQM's involvement in lithium extraction started from a humble project, based on a risky acquisition. In 1983 a public tender took place, for the sale of extraction rights to minerals present on Salar de Atacama.[62] As a result, Corfo gave rights to extract 180,100 tonnes of metallic lithium (multiplying it by a 5.323 conversion factor gets the value in LCE) to Minsal—Sociedad Minera Salar de Atacama, a public-private joint venture between Corfo, Amax Inc. and Molymet.[63]

Molymet these days is the biggest processor in the world of molybdenum, a metal that, due to its stability under extreme temperatures, finds multiple applications across industries.[64] Amax Inc., despite having a great history, lost much of its weight in the metals industries—in the opinion of some, due to excessively wide-ranging diversification.[65]

SQM fully acquired Minsal by 1995,[66] and Ponce wrote to shareholders that the Minsal project had the rights to lithium resources with the largest economic potential in the world.[67] Minsal was not the first of Corfo's efforts to commercialize the lithium in Chilean ground. With Pinochet still in power, as early as 1980 it awarded rights to the salar's exploitation to the Sociedad Chilena del Litio—again a public-private joint venture, this time between Corfo and Foote Minerals,[68] supplier of rare minerals, dating back to 1876.[69] The incorporated entity changed ownership a number of times, ending up under the control of Albemarle, today's largest lithium producer and SQM's only competitor on the Atacama plane.[70]

In fact, Ponce must have thought extensively about lithium before. Perhaps his ideas had been cultivated through conversations with his colleagues from Corfo. This we do not know. What we know for certain, however, is that SQM participated in an auction of rights to a key Bolivian lithium resource, Salar de Uyuni, in the early 90s. The hit to Ponce's ambitions to exploit Uyuni came from an unexpected player, the Bolivian Armed Forces, who lobbied to disqualify the Chilean company with close links to the state from a tender that would award it terrain of 100km^2 only 50km away from the Chile–Bolivia border.[71]

SQM increased its stakes in Minsal step by step. At first, the company were told they could only acquire a part of Minsal if they agreed to leave the lithium exploration to the other investor in the joint venture—FMC,[72] the US company that went on to become the major producer of lithium hydroxide.[73] SQM was offered alternative opportunities in boron and potassium.[74] This condition obviously did not make the company happy, and after negotiations behind closed doors it was dropped. SQM was interested in taking over the whole company, and to achieve this they had to buy out both Amax's and Molymet's shares in the project. Amax at that time was not doing very well financially.[75] FMC, already in the lithium business, also saw huge potential in Atacama lithium and made an offer of $7 million.[76] SQM bid unusually high for this type of transaction at that point in history—$12 million.[77] On the final night of negotiations, an Amax executive called his American colleague from FMC and informed him that he would go for the SQM offer. 'Life is difficult,' he said.[78] With the acquisition of another stake from Molymet, SQM paid in total $18 million to secure access to Atacama's resources,[79] ending up with an 81.82 per cent stake in Minsal.[80] The remainder was in the hands of Corfo—the government— and not much could be done about that.

The fact that SQM had acquired control of Minsal did not mean that they could do as they pleased in Atacama. Lithium was declared a 'strategic mineral' in Chile in 1979, since its isotope is used in nuclear fusion. The regulation remains in force to this day, despite the fact that the nuclear industry now uses only a tiny amount of lithium[81] and that the US stopped treating lithium as a strategic resource for the military industry in 1998.

Chile's laws state that the government has full rights to the lithium in the salars, and they can only temporarily be leased out to companies. No new company has received a concession for the past twenty years, leaving SQM and Albemarle as Chile's sole producers. As soon as SQM bought out Minsal, it had to enter into a lease agreement with Corfo for 81,920 hectares of the terrain, with an expiration date at the end of 2030.[82]

The situation is starkly different for other minerals, where companies can obtain concessions that give them direct ownership rights to the resources in the ground. Another government agency, the Chilean Nuclear Energy Commission (CCHEN) restricts the total amount of lithium that companies can sell through a quota. SQM cannot market more lithium than they set out, no matter the market conditions.[83]

Relations between SQM and Corfo have been difficult for decades, and these dynamics certainly do not benefit Chile. In the opinion of the ex-head of Corfo, Eduardo Bitran: 'I think the most difficult problem now [...] is the big problem with SQM, in terms of the fact that they have been playing complex games in the Chilean political system'.[84] Bitran is suggesting that SQM's involvement in politics has backfired. The long-standing tension between the company and the agency peaked in 2014, when Corfo launched an arbitration process over a dispute regarding the payment of leases.[85] The agency claimed that SQM had failed to pay in full the amount established in the lease agreement.[86]

It might come as a surprise to the outsider that Corfo was not using the arbitration process merely as leverage to get more revenue from SQM. They were in fact trying to cut off the country's only successful producer of lithium from the resource, by ending the lease agreement before the time elapses.[87] The second arbitration proceeding, for which Corfo filed in 2016, proved that they were not joking.[88] Since 1993, SQM has had two contracts in place with Corfo: one for the lease of the salt flat terrain and the second for exploitation of mineral resources.[89] When Corfo realized that the first arbitration, related to the termination of the lease, was not going as smoothly as they had hoped, they tried to attack from another angle, aiming at nullifying the second contract. 'Corfo has notified SQM Salar [SQM's operational unit] that diverse and serious breaches of the obligations established in the contract have taken place, in particular, the obligation regarding the safekeeping and integrity of the mineral properties owned by Corfo'.[90] SQM, the Corfo statement said, 'has breached seriously and repeatedly the existing contracts and is not a reliable partner, that is why [Corfo] has requested the early termination of the contracts to mine the salt flat.[91]'

SQM changed its strategy from defence to attack, this time filing for arbitration from their side. SQM asked the arbitrator to determine whether the payments Corfo claimed SQM owed were based on 'all the facts, and not only from the arbitrarily selected period as Corfo has sought in its previous claims'.[92]

The conflict came to an end in 2018, with the result that SQM was allowed to expand its production by an additional 1.86 million tons LCE through 2030, and expand capacity to 216,000 tons LCE per year.[93] In return, SQM agreed to: make a payment of $17.5 million to Corfo to settle the arbitration, provide $10–15 million to local communities, commit up to 18.9 million to R&D expenditure and make 25 per cent of additional production available preferentially to Chile-based consumers.[94]

The last clause deserves attention, as it relates to Chile's ambitions to move up the value chain, while building a cluster of industries around lithium extraction. The key argument behind this relates to the infamous resource curse, a paradox which holds that countries with abundant natural resources tend to have lower economic growth, less democracy, and worse development outcomes than countries with fewer natural resources. Countries such as Congo and Angola are often quoted as an example. Nigeria, a much richer economy with abundant oil resources, yet still with its fair share of problems, is perhaps more comparable to Chile. It consistently ranks among the world's top ten oil exporters, yet it has to import the majority of the gasoline burned in its own citizens' cars. The reason is that Nigeria does not have adequate oil refining capacity within its borders, as this is a more technologically advanced business than that of pure oil extraction. It is a lost chance for the Nigerian economy, as cheaper fuel would lower both the living costs for its citizens as well as the operational costs for their businesses. It also means that Nigeria loses out on the opportunity to export more added value and thus more expensive products.

It is not only lithium that speaks in favour of establishing a more robust industrial base for the battery and e-mobility sector in Chile. The country also has very low electricity costs that are forecasted to reach $15/MW in 2025, mostly due to the large share of solar energy in the energy supply mix.[95] European electricity costs are on average four times higher.[96] If, based on cost rationale, it pays for almost all London's Uber drivers to transition into EVs, the incentive should be much stronger for Chilean drivers.

Chile is also the world's largest producer of copper—and electric vehicles use about four times more copper than their gasoline-powered counterparts.[97] Added to this, in the future, there should also be demand for copper wiring for the charging stations and related infrastructure needed to support EV growth. Copper foils are also an integral and important part of every battery, used not only in EVs but also in the consumer electronics industry.

There is another side to this argument, however, which holds that this climb up the value chain is pointless and leads to inefficiencies. It is based on one of the most basic and widely known theories in modern economics—David Ricardo's theory of comparative advantage. At its core, this theory says that local economies should focus on developing industries with the ability to produce goods and services at a lower opportunity cost than that of their trade partners. In that way, in an open global economy, all actors, at all times, can mutually benefit from cooperation and voluntary trade. Lower opportunity cost means that you miss very little of something while doing some other activity. Globally, Chile has a high comparative advantage in the lithium extraction industry. By investing in lithium mining, it misses very little, as lithium mining is one of the most attractive investment opportunities in Chile in terms of current and potential future returns. The question that we should pose is how far does—or could—the comparative advantage in the battery supply chain extend for Chile? Does lithium processing on-site also provide the country with a comparative advantage, or should this activity be left for Asian lithium buyers? And what about battery components?

Advocates of sticking to just mining emphasize how difficult it is to produce the high-quality lithium compounds required by top automotive manufacturers. Meanwhile, they argue that Chile is uniquely adept at its mining activities. Chile has earned its unique mining know-how through years of trial and error on how to produce lithium of the highest quality from the brine. Chilean producers started with nothing but a resource. The engineers had to learn from scratch how to even quantify the amounts of lithium and impurities in the brine. Avoidance of possible contamination during the extraction process is another matter. To extract lithium from the brine, you need to apply a certain amount of energy to wake the lithium element up and demix it from the salar, where it lies in a 'dormant' state. The amount of energy and reactants applied does not increase linearly, but closer to exponentially in relation to the amounts of lithium that are being targeted for mining.[98] Despite the process being energy intensive, in a proper environment it is highly ecological. In arid and sunny places such as Atacama, around 70 per cent of the energy needed to recover lithium comes from the sun.[99] In less suitable places, this ratio might be less environmentally favourable. The amplitude of temperatures is also important, as in some processes it is easier to separate sulphate impurities when the temperature is low.[100] Chile's unique environmental setting and the quality of the resource are clear arguments in favour of focusing on mining, if you follow the advice of the Ricardian theory of comparative advantage.

Furthermore, even in places such as Atacama, the lithium recovery efficiency from brine is only around 60 per cent, so there is still much room for improvement.[101] Why not focus on that, instead of building battery factories?

The complexity of achieving battery-grade lithium chemicals is underappreciated, even by some industry insiders. If you want to make 100kg of lithium carbonate from the brine, you need to process around 15,000kg of it.[102] But if you were to add one tablespoon of unpurified brine from the same salar that you produced from to the 100kg of your final product, all the work would be for nothing.[103] Even such a tiny amount would make the material unsellable. This short thought experiment shows how stringent quality requirements are for lithium destined for the battery industry.

Typical battery-grade lithium compound is of 99.5 per cent purity, but this high level is not the most important parameter from a business requirements perspective. Production of high quality battery-grade lithium is, in essence, the art of fine-tuning the levels of impurities in the 0.5 per cent of remaining material.[104]

The 0.5 per cent stands for 5000 ppm, or parts per million. Typically, if your end consumer, a battery cathode material manufacturer, finds more than 100 ppm of calcium, 200 ppm of sulphate or, let us say, a mere 5 ppm of iron in the lithium consignment, he would refuse to accept the delivery.[105] Five ppm is an extremely small amount—it is equivalent to a portion 1,000 times smaller than a teaspoon's worth, thoroughly mixed in one litre of water.

Each end buyer of lithium has different requirements regarding which impurities can be present and at what level.[106] These requirements are often non-negotiable, and directly correspond to the production processes of a specific cathode material maker.[107] So one cathode material maker would require from you, for example, lithium carbonate with a maximum 15 ppm of boron, while for another only 10 ppm would be acceptable.

The advocates of the lithium producing industry argue that, in fact, Chile, unlike Nigeria from our earlier example, does not sell a commodity—a product of extraction—but an added value product, a specialty chemical that in the opinion of some will become more difficult to produce as cathode materials makers tighten their requirements, driven by battery technology advances.[108]

Thus, is it not better to concentrate on improving the efficiency of the extraction processes, and to further push the limits in control over quality, than to try to carve out for Chile a place in the battery industry where players such as Japan and South Korea are already so much more advanced? Besides, every country can potentially build a cell manufacturing plant, but very few countries enjoy quality lithium resources.[109]

In April 2017, Corfo organized a tender. The idea was to entice large, experienced battery manufacturers to set up their factories in Chile, in exchange for supplies of lower-price lithium.[110] Naturally, SQM and Albemarle, the only two lithium producers in the country, had reasons to not be very happy about the arrangement. The national-defence-related regulations that classify lithium as a material needed to make bombs have kept new entrants into the industry almost at zero. Thus, the onus of supporting the development of the Chilean battery sector has been pushed onto the shoulders of these two private players.[111] Since the demand for lithium is constantly growing, SQM and Albemarle need to expand the volume produced if they want to maintain their respective market shares. They cannot do this, however, without the permission of the government. So the government made an allowance for expansion, subject to their allocating a part of their future production to the Chilean battery industry at below market price. We are not talking about a small part of the production either, but a quarter of it.[112]

The tender for the construction of battery production facilities was won by the Chilean Molymet; the Korean companies Posco and Samsung, planning on a joint venture; and the Chinese Sichuan Fulin Transportation Group Co.[113] Among the tender winners, only the Korean players have real experience in the battery industry. The exact plans of all the companies have not been revealed to the public, but are centred around building battery component production facilities—the best guess would be cathode materials, with Samsung possibly mulling over the creation of a battery assembly plant.

To date, Samsung has not built its battery facilities based on where lithium is extracted. It rather has followed demand sources, establishing plants in China and Hungary mainly to serve the German automotive sector.

The Bachelet administration regarded the tender as a success, announcing that the three projects together would bring in $754 million in investment to Chile and at least 664 jobs—many of which would be in higher paid and more attractive engineering roles.[114] Chile's neighbours in the Lithium Triangle, Bolivia and Argentina, also share its vision of building domestic battery supply chains, but in Chile it seemed within touching distance.

It did not take long for the first disillusionment to occur. A year after the results of the tender had been revealed, one of the players—Posco—decided to withdraw. The South Koreans argued that they need lithium hydroxide for their factories, while all Chile could provide in the requested quantity is lithium carbonate.[115] Albemarle does not produce lithium hydroxide in Chile, but it does so elsewhere. SQM produces lithium hydroxide in the country, but in limited quantities. One could question why Posco did not know this beforehand. Of course they did; it is public information, but either they counted on SQM and Albemarle changing their product mix in Chile in line with the market, or they did not expect that the transition to using lithium hydroxide instead of carbonate in cathode materials would be so fast.

After Posco, other tender winners either gave up on their plans of Chilean expansion or did not make any progress.[116] Whether the lack of hydroxide or falling lithium prices or both were the reasons for those players' retreat from Chile is hard to say. The more privy to the lithium industry you become, the more you understand that this is a sector where the future plays an enormous role, perhaps more so than the present. This is probably what makes it so exciting. But the lithium industry sometimes takes one step forward to take two steps back. Much resource, especially in the form of time and money, is being intractably lost on projects that stop in their tracks, never coming to fruition.

SQM had to finance Minsal right after taking over, pumping $170 million into the project.[117] In fact, financing SQM's foray into lithium was the reason the company debuted on New York's stock exchange.[118] The minority stake that Corfo owned was sold to SQM in 1995, as part of a wider movement to offload the state's minority ownership in Chilean companies. Upon the sale, the terms of the contract had been slightly changed, much to the ire of future Corfo presidents. According to the new clause, it was SQM who had the prerogative to extend the contract after 2030.[119] Feeling safer in its position, SQM invested another $275 million over the next three years, building three plants in the Atacama with production capacities for potassium chloride, potassium sulphate, boric acid and, last but not least, lithium carbonate, under the entity SQM Salar.[120] The name 'Minsal' is no longer used.

Since the very beginning, SQM's plan was to be the most cost effective producer in the world. Within two years it had 18,000 tons lithium carbonate production capacity and around 30 per cent of the global lithium market, boasting an already quite diversified global customer base.[121]

In the late 90s the lithium market was much smaller, with global demand at around 20,000 tons in lithium carbonate equivalent.[122] Lithium compounds were predominantly used in ceramics, glass, and primary aluminium production, representing more than 60 per cent of the market.[123] Other major end uses for lithium were in the manufacturing of lubricants and greases, and in the production of synthetic rubber. Even back then, there was already a discussion about new, exciting applications of lithium in electric vehicles. When SQM went on the market with an additional 10,000 tons produced at the lowest cost in the world, it initially brought mayhem upon its competition.[124] SQM, seeing the success of its enterprise, sought to obtain permits for expansion of capacity and extension of the lease for another three decades to 2060.[125] The change of the government, and thus the change of team at Corfo, put an end to these plans. The dynamics between SQM and Corfo have remained pretty much the same over the years since. SQM has been offering increasingly large amounts of cash for the rights to the terrain, expansion of production or the extension of the lease.[126]

Corfo has not seen SQM's behaviour as a fair game. In the eyes of its representatives, it was as if a tenant attempted to take the house away from the landlord.[127] Eduardo Bitran, former vice-president of Corfo, argued that SQM registered its name for Atacama's water rights, when in fact the company is nothing more than just the renter.[128] Allegedly, SQM took every possibility to push for lease extension beyond 2030.[129] It was totally understandable from the view of the company, which had invested enormous amounts of money in development of the salar, but it infuriated people at Corfo. The way they saw it, in 2030 there would be another tender and the lease would go to the highest bidder.[130] SQM, by marking its territory and putting its name instead of Corfo's on anything to do with the Atacama Salt flats, made Corfo's proposition less attractive to any other future bidder, thus skewing the playing field in SQM's favour.[131]

What a foreigner may miss in this salt flat conundrum is the emotional load to the whole affair. Here is a company partially owned by the Chilean ex-dictator's son-in-law, which is accused of breaking contractual agreements with the regulator, supposedly with little regard to the salar's ecosystem. And as if this were not enough, it allegedly takes every opportunity, including 'revolving door' tactics that target the regulator's staff, to make sure the lithium in the salar becomes effectively its own by prolonging the lease to 2060.[132] Rafael Guilisasti was a member of Corfo's board[133] during a legal dispute with SQM, only to be nominated shortly after as a president of Pampa Calichera (Ponce's holding company).[134] Due to his post at Corfo, he was supposed to be familiar with Corfo's negotiation strategy with SQM.[135]

For non-Chileans, the acrimony between SQM and Corfo seems a folly. Would not the Chilean government do better by supporting its world-leading lithium company to the benefit of the nation? It is striking that for many years, SQM paid a royalty to Corfo—6.8 per cent of the value of the goods loaded on shipping vessels—while its American Atacama-based competitor, Rockwood (in 2015 acquired by Albemarle), paid none.[136]

The reasoning was that for many years Corfo had a stake in Rockwood's operation.[137] Nonetheless, like Corfo's stake in Minsal, the Rockwood stake was sold at a time when lithium did not seem such great business.[138]

In recent years, both Albemarle and SQM have faced a sliding royalty regime. It has been a prerequisite, together with allocation of cheaper lithium for Chile's battery sector, to allow for their further expansion. Under a new regime, when the price is less than $4,000 a tonne, royalty is at 6.8 per cent, rising to 10 per cent when the price is between $5,000/t and $6,000/t, and then up to 25 per cent when the price is between $7,000 and $10,000/t, peaking at 40 per cent when the price is above $10,000/t.[139] Critics of the royalty scheme say that with these rates Chile is no longer the lowest-cost producer. Corfo begs to differ. They compared royalty regimes around the world with the cost structure of Atacama-based operations, and came to the conclusion that one needs to look not at average production cost of lithium but at marginal production costs, meaning the change in total production cost that comes from producing one additional metric tonne of lithium carbonate.[140] The argument goes as follows: SQM is already pumping massive volumes of salt brine, to extract potassium from the salar.[141] It could extract more lithium, but it does not—either because this would lead to exceeding the government-set annual quota, or because there is not enough demand anyway. Thus, SQM needs to incur the cost of pumping the unextracted lithium back into the salar, so as not to deplete the resource over time. Estimates are that SQM pumps out half a million metric tonnes of lithium carbonate equivalent per year—more than the world currently consumes— and has to pump most of it straight back.[142] Eduardo Bitran argues that when you treat the costs of pumping and reinjecting the lithium back into the salar as sunken costs, the marginal cost of producing lithium is less than $2,000/ton.[143] From that perspective, SQM's Atacama operation is the lowest-cost producing operation in the world. Meanwhile, this marginal cost approach does not apply to Albemarle, which extracts lithium for lithium's sake, not as a by-product of potassium production.

The implications of this are huge. Let's focus on SQM here, as Albemarle has more facilities outside of Chile. First, the company might actually be more profitable when lithium prices are lower; and second, the price of $10,000/t has been well within a sustainable price range for quality battery-grade products in recent years. There is no reason to set the progressive royalty rate so high, at a relatively low price level. The way the current brackets for royalties are set might be precisely the reason why the two companies have failed to provide cheap lithium hydroxide to potential domestic battery makers. Historically, most of the time, there has been a substantial premium for lithium hydroxide. Thus, not only would the two companies have to give up 25 per cent of their lithium hydroxide production on the cheap domestically, but also, whatever was left for export would face a crippling 'royalty' tax of 40 per cent.

Corfo should be well aware that by pursuing its battles with SQM, they are creating artificial bottlenecks in the development of the Chilean economy. But considering the ex-dictator's son-in-law's level of ownership of and control over SQM, as well as society's sensitivities to past political issues and future environmental concerns, they have to tread carefully. They have often taken the state of negotiations with Albemarle on similar issues as a gauge, to maintain a level of fairness in dealing with both companies.[144]

SQM started to produce lithium carbonate in 1997, but only began making lithium hydroxide—the inadequate supply of which was the reason for the battery components makers' withdrawal—in 2005.[145] It was the development of the battery industry that put pressure on SQM to change and innovate, requiring stronger and stronger quality specifications. In the beginning, nickel cadmium batteries were all the rage, used in electronic devices such as laptops and cellphones. They were quickly replaced by lithium-ion batteries. The main problem with nickel cadmium cells was that they suffered from 'memory effect': if they did not get completely charged after each use, they would potentially charge up only to the last highest charge. This of course is not very practical in the daily hustle, when we charge up our phones, often on the run, just to get 10 per cent more charge before leaving our home or office. Lithium-ion batteries also require less space to store the same amount of energy.

Despite the extremely successful entry of SQM on to the lithium market, lithium did not start out as the company's main revenue maker out of the Atacama Salar. Lithium was originally treated more as a lucrative byproduct of the potassium chloride production process, which at that time fetched historically high prices on the global market. The situation started to change for SQM when the profile of lithium customers changed. Previously, SQM had mainly industrial buyers, more interested in quantities than in tight specifications. Then battery makers appeared on the scene, each of them requiring very specific assays in initially much smaller quantities. The growth of demand from the battery sector was so high that reportedly SQM even started to ship lithium on planes to customers in need, so as not to hold up their production process. Lithium, like other specialty chemicals and commodities, is usually shipped in marine containers, accommodating around 20 tons of lithium per container. From the port terminals used in Antofagasta, Mejillones and Iquique, it takes up to forty-eight days to get a container to Shanghai. By plane it takes one day, but is of course much more expensive.

Argentina, representing another side of the Lithium Triangle, has the second largest lithium resources in the world, estimated at 17 million tonnes.[146] Its natural resources are almost twice the size of Chile's, yet Argentina mined close to three times less lithium than Chile in 2019, also placing it behind domestic mine production in China.[147] Currently Argentina, similarly to Chile, has only two brine-based operations, one belonging to Livent at the ominously named Salar del Hombre Muerto, and one at Salar de Olaroz, operated by Orocobre.[148]

Argentina, unlike Chile or Bolivia, is characterized by a big variety of available salars, which are extremely different from each other. Looking to the future, this may help the local lithium industry, as having different sources of lithium dilutes the risks that exist to the continuity of deliveries and to the success rate of new projects.[149] Another point of differentiation for Argentina is the large number of new mining projects, at present around forty, divided among multiple different operators with many different processes and approaches.

When trying to develop a new project, especially with brine, the worst approach is to just copy it from somewhere else. Resources and climatic conditions tend to be unique, so you have to adjust the design of the extraction process to account for them. For instance, particular elements exist as impurities in different brines. Salars in the Lithium Triangle are also characterized by different weather conditions and microclimates: sometimes it may be raining in the south of a larger salar while in the north the sun is shining brightly. Smart lithium players start with installing weather stations on their properties, to collect precise data for the location before they begin any mine- or process-designing activity. A pond's output can be stable, but the input is variable and depends on the weather conditions. You need higher input volumes when you have higher evaporation rates, and lower input volumes in the reverse situation. You need to understand the weather conditions for a given place beforehand, to be able to plan properly. It is not only about size, or the number of ponds needed. You even need to plan your procurement based on the weather conditions. If the precipitation rates are higher, you will need a larger volume of reactants as inputs.[150]

The production of customized lithium compounds is extremely complex and challenging if looked at in detail, but from a high level, it is very simple to describe. First, lithium-containing salt brines need to be concentrated, a process that takes place through solar evaporation in a series of ponds. The evaporation process produces an intermediate product, lithium chloride, to which soda ash is added to obtain lithium carbonate. This is often where the process ends. If one wants to end up with lithium hydroxide, further addition of lime to lithium carbonate is needed.[151] Significantly, the logistics of getting reactants, such as soda ash or lime, to remote facilities within the salars do not make any of this an easy feat, and by volume considerable quantities of the two are required.[152] Reactants sometimes need to be imported from outside the Lithium Triangle and paid for in dollars, which for Argentina also poses a set of problems.[153]

Argentina is known for a number of things: great steaks, tango, Luis Borges and its predisposition to financial crises. Argentina has defaulted on its sovereign debt eight times since independence in 1816, most recently in 2014. To be clear, default means that the government refuses to pay its debt. Government debt is usually considered one of the safest investments, thus the return on it is usually low. Most countries in the world have never defaulted on their debt, so Argentina's track record clearly stands out on the global stage. Since 2014, things have not really improved in Argentina, despite hopes for change.

When Mauricio Macri won the presidential election in 2015, taking the office from Cristina Fernández de Kirchner who had allowed the financial default, he was a darling of Wall Street. For starters, he was the first non-radical, non-Peronist president in many years. Peronism as a political movement played and continues to play an enormous role in Argentina's political scene. It was founded by Juan and Evita Perón, and popularized by first the musical and then the movie 'Evita'. The Peronists, however, were not necessarily warmly welcomed by the mining industry.

The presidencies of Peronist Néstor Kirchner and then his wife Cristina Fernández de Kirchner lasted from 2003 to 2015. They were characterized by currency controls, so as to maintain an artificially strong peso against the dollar, and generous policies of government subsidies and social spending. After a default in 2001, the Peronists did manage to stimulate the economy, and later on created the illusion that Argentina's economic situation was better than it really was. These policies, however, directly led to another default in 2014.

Macri had a different plan for Argentina: during his first hours in power, he took the currency controls down. He also started to implement a plan for debt restructuring and for cutting government spending. Financial markets reacted enthusiastically. Macri's plan was to bring foreign investment to Argentina, with pro-capitalist, pro-free market reforms after many years of left-wing induced isolation during which Argentina was treated as close to untouchable by foreign investors. Bringing lithium companies to Argentina was a natural fit for that larger plan.

It was not only the change in the administration's attitude and the existence of the resource itself that incentivized junior lithium players. During Macri's rule, restrictions on taking profits out of the country and a burdensome 5 to 10 per cent tax on mining export revenues were eliminated, together with restrictions on imports of mining equipment and parts. The labour market was also deregulated, which opened the way for lithium juniors with tight budgets to utilize flexible hiring policies.

Argentina is a federal system, with each province holding its own unique body of laws; thus, each province has its own regulatory framework for mining. A lot also depends on the governors and mining secretaries of the respective provinces. Three of the most important provinces from the perspective of opportunities in lithium are Salta, Catamarca and Jujuy, where the governors are known for being very supportive of the lithium industry, as it brings jobs and fiscal revenues to their regions.[154] In spite of political changes in the capital, the governors tend to stay on in their roles for surprisingly long periods of time, which is perhaps an indicator that they are not doing a bad job. Juan Manuel Urtubey was the governor of Salta province for twelve years, from 2007 to December of 2019, while Lucía Corpacci has been governor of Catamarca province since 2009. Surprisingly, considering their pro-industry approach, they both belong to the populist Peronist Partido Justicialista. To highlight the importance of the governor's post, it is worth mentioning that when Tesla executives came on business to Argentina, they met with the governor of Salta province.[155]

Much depends on the government on another front as well. Argentina, unlike its copper-rich neighbour Chile, has never been a mining country. This lack of mining tradition can be advantageous—for instance, Argentina has not inherited any cumbersome, outdated laws, and is free to create a legal framework fit for our times, friendly to the lithium mining industry and to the environment at the same time. There are disadvantages too, of course, mainly the lack of infrastructure, as well as uncertainty regarding the role the state will play in support of Argentina's mining ambitions.

To take off, mines need access to gas, electricity, water and railroad connections, in remote locations and at acceptable rates. They require direct connection to ports, ideally through railroads (always a cheaper option than trucks), from which they can ship their product worldwide. It is a two-way street: mining needs infrastructure to prosper, and at the same time supports its development. In successful projects, the infrastructure created can subsequently be used by businesses from other sectors of the economy.

Somewhere in the midterm of Macri's presidency, his popularity started to fall. When his predecessors, the Peronists, were in power, the heavy subsidies to the power sector meant that a monthly electricity bill in Buenos Aires might cost as little as a pack of cigarettes. Macri, on the other hand, in his austerity speeches encouraged his compatriots to face the colder days of the Argentinian winter by turning the heating down and putting on warmer clothes. Argentinians were by then used to financial trouble. It is rare for the middle class Argentinian to trust the country's banks or currency. Savings are hidden at home in dollars, while more well-off or financially savvy citizens keep their money in offshore bank accounts. This does not mean that they were not fed up with the situation in the country. They knew very well, in electing Macri, that he was going to tighten the belt. As long as the measures brought an improvement to the economy, they were ready to go with them. But inflation remained very high, at over 20 per cent in 2017, over 30 per cent in 2018, and hitting over 50 per cent in 2019. To rescue the peso, Macri took a loan from the IMF, to the tune of $50 billion, its largest ever awarded. That was enough for most Argentinians. Daily life became more difficult under Macri, and at the same time his austerity model failed to deliver the expected fixes to the economy, instead putting it further into debt. In the primaries in 2019, to the surprise of Wall Street, Macri received only 32.1 per cent of the vote, while Alberto Fernández (a Peronist) got 47.7 per cent. Markets reacted immediately with the sell-off of Argentina's assets, only inducing the peso to fall further.

Foreign investors, well aware of the Peronists' history of industrial nationalization and currency controls, hindering the repatriation of returns on invested capital and increasing the cost of doing business, started to withdraw from the country. The worst-case scenario materialized when in 2019 Fernández won the election. Meanwhile, the situation had become so dire that it was the pro-free-market Macri who, before his departure, re-implemented currency controls, which were then, unsurprisingly, maintained by his Peronist successor.

Some argue that the situation is not that bad for Argentina's lithium industry. Fernández even met with lithium miners before assuming the presidency, reassuring them of his unwavering support.[156] There are practical reasons for that. Lithium exports generate hard currency, which is key to the repayment of the country's US dollar denominated debt. The currency controls are the major problem, though. When companies receive payments in US dollars, they are bound by law to quickly exchange them into pesos. They need to do this at an official exchange rate, which is much lower than the black market rate. Meanwhile, a large share of the domestic contractors and employees demand payments for goods and services that reflect prices at unpegged, black market exchange rates. When the company wants to buy inputs from abroad, be it reagents or machinery, it needs to exchange its pesos to dollars, while the spread between peso buying and peso selling prices is very wide at the official exchange rate. Thus, the company loses on exchanging dollars to pesos on exports, and pesos to dollars on imports, when, in an ideal situation, it would just hold the dollars in its bank accounts and wait to use them. That said, what is disadvantageous to single companies such as these is advantageous for the Argentinian currency, as it increases its strength.

Mining companies tend to be resilient to political and macro environments. You dig where the resources are, and too often they happen to be in emerging or frontier markets. Large multinational companies have been successfully mining for cobalt in Congo, despite part of the country being at war. Argentina's economic problems seem small in comparison. Nevertheless, the lithium industry in Argentina has already experienced its first major setback.

Eramet is a large French mining company, with a history dating back to 1880. It was funded by the legendary Rothschild banking family, probably the wealthiest family in modern history and the subject of many conspiracy theories, whose prominence peaked in the nineteenth century.[157]

Nowadays Eramet is a stock listed company, reinventing itself as a battery metals provider.[158] Its work on the Centenario lithium project in Salta has been halted, despite being at an advanced stage, with hundreds of millions of dollars already invested in the project.[159] The company announced that concern about Argentina's macro-economic environment was one of the main reasons behind the decision to stop operations.[160] Time will tell if other projects will follow Eramet's example.

Despite the tough measures on the macro-level, taken to save the country's economy but impacting all industries within its borders, there are some very favourable mining-specific tax breaks and royalties available in Argentina. While Chile levies up to 40 per cent royalty on the final product, Argentina takes only a flat 3 per cent royalty at a pithead price.[161] There are also no municipal taxes and stamp duties on lithium projects. Provincial governments do what they can to attract new investment in lithium.

Their eagerness comes at a cost, namely in the criticism the lithium industry attracts. All extractive industries are invasive by nature and leave a mark on the environment. With as many as forty different lithium projects in the country,[162] the potential impact on the environment is very high. The main problem is that lithium mining is very water intensive and takes place in already extremely arid areas, where water is a scarce resource. Salt flats are unique places on earth, formed by thousands of years of evaporation of water bodies such as lakes or ponds, due to strong exposure to the sun (Atacama has 30 per cent more sun exposure than the Mojave desert). Bolivia's Salar de Uyuni, under which the largest lithium resource in the world[163] can be found, is a result of the evaporation of several prehistoric lakes that are between 30,000 and 40,000 years old, and has a depth up to 140 metres. Salars are also flat to a seemingly otherworldly degree—on the 10,260km^2 of Salar de Uyuni the terrain changes by less than a metre in height, which makes it ideal for calibrating satellites.

Salars, despite their unwelcoming appearance, are host to diverse ecosystems. They are also surrounded by ethnic groups who have inhabited these areas for centuries, and have adapted to making the best use of what nature has offered them.

The question is whether these local communities benefit from the exploration of the new oil, and whether all is being done to preserve the salars in the state in which they existed before the mining industry took hold.

It takes around 2 million litres of water to produce a tonne of lithium from a salar.[164] The pumping stations are scattered around the lithium ponds and use kilometres of hoses to bring the brine—a salty, liquid mud—to the evaporation ponds. It is little wonder that lithium mining from brine is sometimes called water mining. Pumps are working at rates as high as 2,000 litres of brine per second.[165]

The fact that lithium consumes a lot of water in areas where it is scarce is well known. What is not known for certain is the impact of lithium mining on unsalted water in and nearby the salars. One theory, preferred by the industry, says that underground aquifers containing brine (salty water) are not connected in any way with the unsalted underground water aquifers that plants, animals and people consume. Thus, it is largely fine to exploit the brine as this does not harm anybody, with maybe the exception of the landscape. The competing theory, believed by some of the scientists and NGOs, is that pumping the brines violates the delicate water balance underground, as clear water replaces the water depleted from the brine wells.

The biggest problem is that there is not enough data to say who is right, especially in Argentina, where the government has left the matter of water surveys in the salars to the very same companies who are accused of their depletion. There are no independent or government conducted studies available to use as a benchmark against the measurements that mining companies are presenting.

According to the law, lithium miners are obliged to conduct EIA (environmental impact assessment) studies before they start building a mine, the results of which require approval by the authorities. The same authorities, however, are also responsible for both the promotion of mining and the approval of environmental studies, which creates a clear conflict of interest.[166]

Personal testimonies collected by NGOs among indigenous communities in the vicinity of lithium extraction sites tell a story of noticeable decreases of available water resources for livestock and human consumption, water contamination, and mysterious deaths of herd animals.[167]

Argentine legislation recognizes indigenous people's rights to develop and control their lands, territories and resources. The country is a signatory of the UN Declaration on the Rights of Indigenous People (UNDRIP), which says that Argentina should 'obtain [indigenous people's] free and informed consent prior to the approval of any project affecting their lands or territories and other resources, particularly in connection with the development, utilization or exploitation of mineral, water or other resources'.[168] Since this is not further regulated through domestic law, the indigenous community approval process is customarily a part of the environmental impact assessment.

The lithium projects that went ahead have been approved by the heads of the indigenous communities, in the case of the Olaroz Cauchari Salt Flats by Huancar and Pastos Chicos.[169] Yet despite the reassurances of the companies operating in the salar, the approval process and the benefits provided in exchange for access to the resource providing hundreds of millions in revenue per year raise questions.

Interviews conducted by Fundación Ambiente y Recursos Naturales (FARN), an Argentinian NGO, reveal that locals have not really felt that their consent was truly informed. Mining company representatives who met with them have not adequately highlighted or discussed the possible environmental risks of the projects. When locals raised these questions, answers came in technical lingo that they did not understand. No third party representatives were present during the meetings with miners, neither from government nor from academia. Indigenous community members did not feel that there was anybody to support them who didn't have a vested interest in the project going ahead. Thus, they never felt like they really had much negotiation power. They were mostly happy to take whatever the companies promised, assuming a rather passive position in the whole process. Jobs were the most important benefit being offered, and many are still happy for the opportunity to work in a place where employment opportunities are rare. One community received a company sponsored secondary school, and some of the members who had previously left to provide better opportunities for their children were able to come back.[170]

The annual compensations the local communities received were in the range of between $25,000 and $60,000 per community, equivalent to the revenue from sales of 3 to 10 tons of lithium carbonate, while the production of one of the companies present in the salar achieved over 12,600 tons of lithium carbonate in 2019.

It does not have to be this way. There are many examples of mining towns around the world that have profited tremendously from a nearby mine, with wealth redistribution through taxation and higher than average wages. Saudi Arabia could bring its indigenous nomadic tribes from rags to riches. Its legendary Saudi oil minister, Ali al-Naimi, spent his childhood in a nomadic tent.[171] Then why is it that countries such as Argentina cannot provide its indigenous communities with middle-class living standards, on the wealth of resources on land their forefathers have lived off for centuries?

The situation is very much different in Bolivia, the third and the most resource-rich country of the Lithium Triangle, which for thirteen years has been ruled by its first indigenous president. The very specific circumstances surrounding lithium in Bolivia, and its long, fruitless history to commercialize the largest lithium deposits on Planet Earth, call for a separate chapter—and it is there that we turn next.

4

SAUDI ARABIA OF LITHIUM

On the night of 11 November 2019, a Mexican governmental plane took off from the central Bolivian town of Chimore, carrying Bolivia's leader of thirteen years into exile. As the aircraft ascended, it left a country in turmoil. The city of La Paz, 'the peace' in Spanish, one of the country's two capitals, had been rocked by violent protests. Employees of most embassies based in the city had been told to work from home. La Paz's central square, Murillo Square, had been turned into a makeshift stronghold, reminiscent of Ukraine's Maidan of 2014.

The protests lasted for weeks since the election, which observers from the Organization of American States deemed corrupted. But not everybody wanted Evo Morales out. He was the first indigenous president in a country where around 60 per cent of the population declare themselves to be indigenous, the highest ratio in Latin America.[1] During his long years in power, Morales substantially lowered the poverty rate and steered the country through years of strong economic growth, supported by a thriving commodities market. In fact, the police, who stopped acknowledging presidential authority days before he left the country, were worried about the masses of his supporters marching towards the capital.

Evo, as he tends to be called, described the end of his reign as a coup. He was not held at the barrel, nor was his brand-new presidential residence—'Casa Grande del Pueblo', made from steel and glass and towering upon the Lapaz landscape—threatened with bombardment. In light of the events on the streets, though, he received a suggestion during a broadcast speech by General Williams Kaliman, head of Bolivia's armed forces, to resign.[2]

In a country infamous for the highest number of coups d'état on the record, to the unbiased observer, following that suggestion might seem prudent. Not everybody, however, was convinced that it really was a coup. The Brazilian president Jair Bolsonaro told Brazil's *O Globo* newspaper, while commenting on the events, 'The word coup is used a lot when the left loses. When they win, it's legitimate. When they lose, it's a coup'.[3]

Some weeks after, when the tensions on the streets subsided and when Evo Morales gave his first longer interview as an ex head of state, he said that he was absolutely convinced that the coup had been about lithium. 'It was a national and international coup d'état. Industrialized countries don't want competition,' he claimed.[4] He went on to state that he had not been forgiven by Washington for having chosen Chinese over American support for the project, and that Bolivia was on its way to dictate the lithium price in the world based on the sheer scale of its resources.[5]

It is a fact that Bolivia has the largest resources of lithium in the world. They have been estimated by the US Geological Survey (a government agency) at 21 million tons.[6] Bolivians maintain that they are much larger yet, estimating that their largest salt flat alone—Salar de Uyuni—holds 140 million tons of lithium.[7]

It is far from the truth though that Bolivia was in any way close or on its way to determining global lithium prices. The most recent data Bolivia's customs published shows that Bolivia exported a mere 20 tons of lithium carbonate in 2018, just enough to fill one container. All of it went to China.

Nevertheless, it is hard to find a country whose national sentiments and strategy have been more defined by the idea of lithium as the new oil. The dream of using lithium to enrich the nation went back to the beginning of the Morales presidency in 2006, and became one of its defining elements. It was also something more to the Bolivians than just a way to boost the country's Gross Domestic Product. Bolivia has a natural resource trauma, which Morales tried to overcome. When Richard M. Auty in his seminal book *Sustaining Development in Mineral Economies* (1993) coined the term 'resource curse' to describe the countries where underground treasures brought less rather than more good, he used Bolivia as one of the examples.

The Potosí region, where lithium-bearing Salar de Uyuni is situated, was the Spanish Empire's greatest source of funding between the invasion in 1545 and the country's independence in 1825. While its silver rich Cerro Rico ('Rich Mountain' in Spanish) has been inflating prices on the Iberian Peninsula, it took eight million mostly indigenous lives in Bolivia. It also raised Potosí to the rank of a global city—in the early seventeenth century surpassing the size of both London and Milan. 'I am rich Potosí, treasure of the world, king of all mountains and envy of kings', read the city's coat of arms. But the importance of Potosí did not last, and ended along with its silver.[8]

After gaining independence from Spain, the power in the country has been held by 'criollos', direct descendants of the Spanish colonists. They often continued to derive their wealth from mineral resources. Families of the 'tin barons', the Patiños, Aramayos and Hochschilds, were, at the peak of their fortunes in the eighteenth and nineteenth centuries, controlling around 30 per cent of the world's tin market.[9] In 1952, their tin mines were nationalized, bringing a solid stream of revenues to the state, until the price of tin crashed by over a half in 1986.

In 2003, the country was rocked by conflict over the exploration of Bolivia's gas resources, the second largest in South America after Venezuela's, leading to the so-called 'Gas War'. This cost Gonzalo Sánchez de Lozada, or 'Goni', the presidency, and opened the door to Evo Morales' victory in the following elections. Sixty people from El Alto, Morales' electoral stronghold, died in protests, outraged by Goni's government's gas deals with British and Spanish oil majors. There were estimations that Bolivia was to get only $40 to $70 million per year from gas exports. To add fuel to the fire, as it were, the gas was to be exported in its rawest form through the pipeline and liquified and shipped to Mexico and US on Chilean territory. It yet again reminded Bolivians of the long history of exports of only the crudest form of commodities and of the bitter taste of losing access to the ocean in the war with Chile in 1884. This was a blow to the collective self-consciousness which annual 'Day of the Sea' marches in Bolivian cities on 23 March make difficult to forget.

Evo used the gas wars as a trampoline to the presidency: he was one of the leading figures in the protests demanding full nationalization of hydrocarbon resources. Signing a decree demanding that all gas reserves be nationalized was one of the first things he did as Bolivia's president. 'The state recovers ownership, possession and total and absolute control' of hydrocarbons, it said.[10]

In 2019, revenues from gas exports brought around $2 billion to state coffers, a far cry from the estimated $40–70 million per year from royalties if the Bolivian gas sector had been private.[11]

Bolivia's state led development of gas fields has been a success. The same cannot be said about lithium, which was equally important on the agenda from the very beginning. In fact, Evo's presidency has been wrongly perceived as a window of opportunity by foreign corporations, whose emissaries started to court the freshly minted president for lithium related contracts and exploration rights. But before we get into that, we should consider the beginnings of the lithium story in Bolivia.

Already in 1974, the Bolivian government declared the lithium rich Salar de Uyuni a 'fiscal reserve', which basically assigned to the state basic ownership of the Salar as well as the legal rights to exploit and administer all of the mineral resources within its boundaries.[12] This happened without any consideration of the rights of indigenous communities who had lived around the salar for centuries. For a long time, there was no real interest in Bolivia's lithium. In the late 80s, the negotiations between FMC-LITHCO, a US corporation, and the Bolivian government started. It took five years until the deal for joint exploration of lithium from Salar de Uyuni for a period of forty years was hammered out.[13] The contract required a signature from Bolivia's president, at that time Jaime Paz Zamora, as well as approval in national congress.

In the early 90s, the demand for lithium was modest, but new exciting battery applications were already on the horizon. General Motors made its first announcement regarding the possibility of a battery powered electric vehicle, and France seemed to be moving on with its plans to build new reactors using nuclear fusion, which would require lots of lithium in the process. FMC-LITHCO, one of the lithium major players, looked to secure a mineral resource base to capitalize on those plans.

FMC-LITHCO planned to move cautiously in Bolivia, as the contract called for an initial investment of $6 million and a three-year feasibility study. If the study proved the exploration from the salar to be economically feasible, an additional $40 million for a processing plant was to be invested. Americans estimated at the time that the project would create 200 jobs and result in expected $100 million in exports over ten years.[14]

From the very beginning, the project was politicized. There were even hunger protests held against it, with banners reading 'Death to those who would hand over the Uyuni salt flat'.[15] The president was in favour of working with the Americans, calling for moderation and a 'modern mental attitude, without complexes or fear of progress', to 'create the best conditions to capture investment'.[16] However, he faced strong opposition in FRUTCAS, for decades an important player in the region and later on Morales' strong ally.[17] Federación Regional Unica de Trabajadores Campesinos del Altiplano Sur, as it reads in full, is an organization representing 40,000 agriculturalists, a large percentage of whom are of indigenous descent from the Salar de Uyuni region.

The contract was signed, but congress had difficulties in accepting it. Social pressures were one thing, but congressmen were also concerned with the wording of the agreement. The contract duration was for forty years, but there was no guarantee that FMC-LITHCO would extract even a gram of lithium during that time.[18] Even if such a possibility was remote, it was a valid concern. If the feasibility study had determined that the project was un-economic, the Americans could have waited for the market environment to change, while blocking access to the salar for other companies.

FMC, on the other hand, was not willing to accept a change in the percentage of value added tax that had grown from 10 per cent to 13 per cent, right in the middle of the negotiation process.[19] They thought that since they had signed the contract in February and the value added tax changed in May, they should not pay the new taxes. This line of reasoning was also unacceptable for the congressmen. Finally, FRUTCAS lobbied against ratification, as, in the hands of the FMC, Salar de Uyuni was projected to leave only 2 per cent of revenue in the region.[20]

In the end the long negotiation and public attitudes discouraged FMC, which left Bolivia in 1993 and went on to establish a very successful lithium mining operation in Argentina's Salar de Hombre Muerto. It took only three years for FMC's Argentinian project to start producing on a commercial scale in 1997.[21] Despite left-wing, nationalist Peronists in power, the project was supported by the government. Technological challenges were also less demanding, with much lower amounts of magnesium impurity in the salar and higher evaporation rates than in Bolivia.

Between 1993 and Morales' election in 2006, Bolivia's lithium resources somehow disappeared from the public agenda. The six presidents in the time between Zamora and Morales had different priorities in raw material space, preferring to focus on the development of the country's gas resources. Their short terms in power and generally turbulent changes in Bolivia's politics stood in the way of the efficient development of projects with longer term horizons.

The first substantial change to the status quo that Morales introduced was a constitutional change defining Bolivia as a plurinational state that recognizes the place of indigenous communities in it. This went together with the active promotion of 'vivir bien'—an alternative concept of development for the country, in harmony with nature, and derived from indigenous communities' ancestral philosophy. Bolivia tried to brand itself as pro-environment, but did it on its own peculiar terms. In 2009, Morales rocked the Copenhagen climate change summit with a speech in which he called capitalism the root of climate change, and proposed drastic, but in reality unworkable measures to stop it. He also called for rich countries to pay climate change reparations to the South, and for the establishment of a climate court of justice.[22] Seeing the limited impact of his proposal, he proceeded to organize his own climate summit in 2010, boycotting existing international structures dealing with the problem.

An anti-capitalist stance, care for environment and pro-indigenous policies defined his official approach to lithium mining.

But according to his domestic critics, his dealings were double-faced and, through the rhetoric of racial distinctions, polarized the nation.[23] The new mining law introduced in 2014 serves as a good example of the difference between Morales' self-branding activities and his actions. The act privatized water rights to mining operators, restricted the process of consultations by affected communities, and, perhaps most abhorrently, criminalized any protest activities against mining.[24]

In the end his approach has been no different from the approach of any other right-wing or left-wing authoritarian leader, to whom benevolent ideology is only a convenient, selectively applied tool to push forward the interests of his supporters.

It is perhaps counterintuitive that Morales, with his allergy to capitalism and rich nations, was courted so much by Western governments and corporations interested in lithium extraction at the beginning of his presidency. It is highly unlikely that government officials or corporate officers did not know about his views. Probably nobody believed that Morales would be wilful enough to go it alone on such a complex project.

Lithium mining is much more demanding than mining for other metals such as silver or tin. Moreover, Morales was not interested in just mining raw lithium material and shipping it abroad for further processing. That would be against everything that he believed in. Bolivia has for too long been just an exporter of raw materials, at the mercy of the boom-bust cycle typical for commodities. He wanted lithium to be processed into battery-quality chemicals, and then used in the production of Bolivia's batteries, and one day even Bolivian electric cars.[25]

To achieve this in a developing country such as Bolivia would necessitate the support of more industrialized nations. It was difficult for anybody with at least an elementary grasp of the topic to imagine that Morales and his cadres would have believed otherwise.

Nevertheless, corporations and governments in Japan, South Korea, France, Canada and Brazil courted Morales' power structures for a long time, and ultimately unsuccessfully.

In February 2009, Bolivia's National Mining Director Freddy Beltran announced that four companies—Japan's Mitsubishi and Sumitomo, France's Bolloré, and Korea's LG—expressed interest in the Uyuni deposit.[26] Meanwhile a lithium producing pilot plant was on the way, developed independently by the Bolivian state mining company COMIBOL. What is significant is that none of the major lithium producing companies at the time expressed any interest in Bolivia's resources. They were likely concerned with the threat of the asset's nationalization, poor infrastructure, and the quality of the resource. This is not to mention the climate, as the heavy rains could deal a blow to any fledgling conventionally producing operation.

To realize the integrated lithium-ion battery economy vision in Bolivia, it was not only enormous greenfield investments that were needed. The problem was also in the quality of Bolivian technical education, and the number of skilled workforce required on the ground. While Bolivia had mining professionals, the same could not be said of chemical and high-tech engineering talent.

Morales' visit to France and his meeting with Sarkozy in 2009 were to a large extent determined by the talks on Bolivia's gas and lithium resources. Discussions on gas were led by high-ranking representatives of Total, and on lithium by Bolloré group.[27] Bolloré is in the business of logistics, holding a strong position on the African continent, and it has been a promising player in Europe's battery market. During the meeting, Morales recognized the need for international investment, but made it clear that Bolivia 'will not sell the resource for any price'. After the meeting, Bolloré announced that it would present its proposal to the Bolivian government regarding the investigation, production, marketing, and even the fabrication of lithium-battery fueled cars in Bolivia.[28]

The meeting's atmosphere was characterized by distrust. Morales told the French companies to 'Be partners not predators' and trumped up his good relationships with Russia, as a card up his sleeve, especially in relation to French-Bolivian gas projects. He made veiled threats, that either Total would meet his expectations in regard to the level of investments and timelines, or Bolivia would be bound to choose what was best for it.[29]

Despite the not entirely reassuring ambience, Bolloré made good on its promise, teamed up with the French metal producer Eramet, and presented a comprehensive development plan to the Bolivian government.[30] It waited in vain for a positive outcome of the negotiations, meanwhile becoming increasingly interested in mining for lithium in Argentina, as hopes for a successful outcome with Bolivia dwindled with time. In February of 2010, the companies finalized an agreement to explore for lithium in Argentina.[31] The fruit of this strategic reorientation was Eramet's Centenario project, which, in spite of shelving amid low lithium prices in 2020, still holds some promise.

The one mining company that was already at home in Bolivia was also interested in developing the Uyuni salt flat. Yet their experiences with the country made their outlook a little less rosy from the start.[32] The Japanese trading house Sumitomo acquired the San Cristobal silver, lead and zinc mine in 2008, amid low metal prices. Sumitomo is a sogo shosha, a typically Japanese phenomenon. Sogo shoshas are general trading houses, known for being active in a wide range of markets as intermediaries. Their beginnings can be traced back to the opening of Japan in the mid nineteenth century, when they acted as economic bridges to the outside world. Later on, they maintained their strong position on the market due to a vast network of relationships around the globe as well as large amounts of hard cash, accumulated throughout Japan's boom years, which ended abruptly in the beginning of the 90s. Subsequently sogo shoshas maintained their position on the market by deploying saved cash in projects in need of investment and in distribution channels around the world, often in the natural resource sector.

One of Sumitomo's ex-employees described the harrowing experience of dealing with Bolivia's state structures at the time. He described meetings with government officials as frustratingly long and frequent, with the officials themselves often highly inexperienced.[33] 'They don't even know what Sumitomo is. I call for a meeting, and they say, "Who?" I get letters addressed to Mr. Sumitomo and calls from officials who want to speak to him. With the Bolivian government, private companies are very low ranking. he added. 'They want to deal with other government officials, not with executives, even though Sumitomo is now one of Bolivia's largest investors.'[34]

Sumitomo also had to confront the negative opinion of officials, who believed that the company was trying to take advantage of San Cristobal's previous owners by buying the mine on the cheap. There was a prevalent opinion that San Cristobal should have been nationalized instead. Bolivians did not understand that Sumitomo took on an enormous risks while buying the mine, assuming its $400 million debt.[35]

It was not only dealings at high level that were problematic— on the ground, too, Sumitomo faced protests and strikes that led to temporary halts in the mine's operations. Japanese executives were also complaining of the high taxation burden the company was facing in Bolivia, and of problems with issues as elementary as VAT refunds.[36]

In the end Bolivia's red tape and always looming risk of nationalization discouraged the company from any involvement in Uyuni. Instead, Sumitomo went on to become a very successful player in cathode materials (more processed materials for batteries, containing lithium) and a main Tesla supplier, instead of being directly engaged in harvesting lithium from the ground.[37]

It also became increasingly clear to foreign players that Bolivia is not just looking for fiscal revenues from lithium in the form of royalties and taxes—what pretty much represents the standard way of doing business in the mining world. Bolivians were only interested in technical support for developing the project, and at the very best would grant a minority stake in lithium producing operations.

By going alone, Bolivia lost its window of opportunity, as the companies willing to spend time as well as tremendous amounts of capital on what always was a risky project invested in other countries or in other parts of the battery supply chain. As Juan Carlos Zuleta, perhaps Bolivia's longest standing lithium analyst, pointed out in a *New York Times* interview at that time: 'We have the most magnificent lithium reserves on the planet, but if we don't step into the race now, we will lose this chance. The market will find other solutions for the world's battery needs.'[38]

While Morales had been testing the waters with foreigners, the real control of the project went into the hands of the newly minted Scientific Committee, a shopfront of FRUTCAS, the farmers organization and staunch support base for Morales.[39] FRUTCAS's chief was a quinoa grower, living by the salt flat.[40] Not surprisingly, the Committee did not make any substantial progress, and, when results failed to materialize, its role was taken over by the state.

Within three years, the state mining company, COMIBOL, managed to establish a lithium carbonate pilot plant at Salar de Uyuni at a cost of $20 million.[41] A pilot plant may sound proud, but it was nothing but a small system to find out more about the behaviour of the process before applying it on a commercial scale. It took a further three years to export the first several tons of lithium carbonate to China. That means that whatever the pilot had been outputting for the first years was probably of too low quality to show the world. Pilot plants have been Bolivia's low-hanging fruit, built to demonstrate to the nation that progress has been made. This is likely why the battery pilot plant and cathode materials pilot plant followed. As per 2019, the battery pilot plant produced only 110 kWh of batteries, equivalent to what is needed to power just two modern electric vehicles, while the cathode material pilot plant brought only 28.5kg of NMC cathode material, an amount more adequate for a university lab.[42]

Bolivia, despite the reassurance of going it alone, relied to a great extent on foreign companies to build the pilots. Neither did the country spare money on the process. Bolivia's largest state investment to date has been the $1 billion that went into gas extraction. The investment paid off, and the gas proceeds were allocated to lithium development. In 2015, Morales pledged to invest $995 million in the lithium industry by 2019.[43] In mid-2019, it was revealed that around $600 million had already been spent on the project.

The investment had little in the way of results to show for it. For the sake of comparison, Orocobre's Salar de Olaroz facility in neighbouring Argentina had been completed at the expense of $229 million and had a nameplate capacity, at the first stage, of 17,500 tons of lithium carbonate per year.[44] Orocobre's project suffered its fair share of challenges and setbacks,[45] but still it serves as an example of the superiority of private firms over state-owned firms in the execution of lithium projects in Latin America.

Bolivia's decision makers' choices in regard to lithium industrialization were dubious from the very beginning. The foreign companies selected to do the work were not really established in lithium and battery space. They were solid companies, but with more experience in adjacent areas rather than in the lithium or battery industry per se. For the construction of a cathode materials pilot plant, instead of teaming up with largest cathode materials producers and major cathode materials plant equipment providers, Bolivia decided to hire ECM Greentech, a French firm that calls itself on its website a 'PV equipment manufacturer'.[46] In short, the company specializes in production lines for solar panels. It seems that Evo Morales' visits and political sympathies played a bigger role in the selection of partners than knowledge of the industry itself. Linyi Gelón New Battery Materials, which was commissioned to build a battery plant pilot, also ranks very far from the top, which is occupied by the largest and most technologically advanced Chinese battery companies. It was yet again a strange choice, considering the governmental level of the negotiations and the amounts of funds allocated. The Bolivian state could have easily worked with the crème de la crème of the lithium economy. For unknown reasons it chose not to.

As the negotiations with different parties proceeded, the land-scape of the lithium industry started to change. This at least was one thing that did not escape the attention of Bolivia's state players. On the market, for EV battery applications, battery-grade lithium hydroxide became increasingly preferable to lith-ium carbonate. Some analysts predicted that hydroxide, as a superior material, would gradually push carbonate out of the market. Bolivia started to look for a partner to establish a lithium hydroxide plant. With the passage of time, Bolivia's expectations also changed: there was a slow realization that it was not enough just to rely on the technical support of foreign companies, and that it needed to invest its own money. Partly it was a lesson from the past, partly fear of the consequences of continuous mis-spending, and partly a negative change in gas revenues, which limited Bolivia's imagination and spending largesse.

Bolivia started to look for a partner and investor who was still willing to accept a controlling share of the state in the project. German private companies interested in Bolivia's lithium well understood the weight that Evo Morales' cadres assigned to rela-tions on an inter-governmental level. At the same time, the German government was aware of the opportunities that Bolivia represents for its firms. Siemens already capitalized on Bolivia's gas boom, by building three gas-powered power plants in the country. The German political class saw an opportunity to provide a raw material for Germany's EV sector, which was getting up to speed with rising interest from the nation's main automakers. From a larger perspective, they were also trying to limit Germany's and EU reliance on the Chinese lithium supply chain. Over the course of two years, Berlin led an intense lobbying effort on a political level. It included several diplomatic visits during which the advantages of moving forward with German companies were pitched. Among the main arguments were the financial backing of the project in the form of government guarantees and an access to Germany's famed automotive sector. Bolivia's officials were also invited to tour German factories, in order to experience German industrial prowess first-hand. To put a cherry on top, the German Minister for Economic Affairs and Energy Peter Altmaier wrote a letter to Morales, emphasizing German firms' commitment to environmentally friendly processes, playing on Morales' 'vivir bien' ideology.[47]

In an area of diplomatic effort, Germany has been facing a formidable contender. Bolivia's relations with China had been at their strongest before the Morales fall. Evo made four official visits to China during his term and perceived Chinese leaders as ideologically close. China has been a buyer for Bolivia's metals and agricultural products and a seller for cutting-edge technologies. China launched Bolivia's first satellite at a cost of $300 million—$251.1 million of which was financed with a low-interest loan from the China Development Bank, payable over fifteen years. Bolivia bought from China a state-of-the art biometric surveillance system, amid controversies that it might be used to prosecute political opponents. China also saw a larger place for Bolivia in the future in its Belt and Road Initiative.

Nevertheless, it was ACI Systems, a family-run, medium-sized company in cleantech and industrial equipment manufacturing situated in the vicinity of Lake Constance in South West

Germany, who won a contract to change the fate of Bolivia's lithium.[48] In an interview, a Bolivian official claimed that a deciding factor in the company's choice was ACI System's willingness to accept the terms—ownership of the salar remains Bolivian, 51 per cent of the stake in the venture is under Bolivian ownership and control, as well as the prospect of a plug-in into German automotive brands' future lithium demand.

The company also pledged to invest up to $1.3 billion in the project. This is an unfathomable amount, as this would probably not be possible for an entity of that size without strong support from the German government. The contract, planned with a seventy-year timeline, was signed by the two companies in Berlin on 12 December 2018 in the presence of Bolivia's minister for foreign affairs and Germany's minister for economic affairs and energy, Peter Altmaier.[49]

The ACI Systems joint venture was supposed to provide 30,000 to 40,000 tons of lithium-hydroxide per year by 2022.[50]

China was not left empty handed, a fact that, unlike the German deal, pretty much escaped the Western media. Xinjiang TBEA Group entered a joint venture with Bolivia's state-owned lithium company on similar terms to ACI, receiving a 49 per cent stake just two months after the Germans.

The Chinese got into two other smaller lithium-bearing salars, Coipasa and Pastos Grandes, and pledged an investment of $2.3 billion. ACI had also been bidding for two smaller salars. Another competitor was Uranium One, affiliated with the Russian state-owned Rosatom, a nuclear energy giant. The Chinese ambassador to Bolivia, Liang Yu, praised the agreement for its 'historical' significance.[51] In what constitutes a pattern here, Xinjiang TBEA Group is not a lithium company either. It is a conglomerate active in the widely defined power sector, perhaps most known for its power plants investments.

The events took a dramatic turn in this story in November 2019, not even a year after the German-Bolivian cooperation had been feted in Berlin. On 3 November, Morales repealed Decree 3738, which had established the Bolivian-German joint venture, through another presidential decree. The planned seventy years of cooperation were crossed out, without any reason given.[52] The managing director of ACI Systems reportedly learned of the news from the radio.[53]

The decision was made on the wave of protests in Potosí. Yet again locals were not satisfied with the 'split of spoils'—the share of the royalties that they will be getting. The greed of local power brokers from Potosí's Civil Committee (COMCIPO) inspired demonstrations that derailed another project on the way to materializing Bolivia's lithium dreams.[54]

The German politicians involved in the deal, up to the Ministry of Economy, which expressed 'surprise and regret' at the event, were inconsolable. The company was also unable to accept the verdict and continued to work on the project from its Baden-Württemberg headquarters, in the hope that all was one big misunderstanding and that the challenges could be accommodated.[55] ACI Systems likely did not know about the demands by the local opposition and the prevailing lack of satisfaction about the profit sharing in the municipality. The Bolivian state probably took on any dealings with the local stakeholders themselves, unwilling to share their concerns with the Germans so as not to discourage them from the investment.[56]

The German ambassador to La Paz tweeted, 'Germany once again confirms its commitments to Bolivia in the lithium project and keeps its word. And Bolivia? Stopping the project would be a hard blow to our bilateral economic relations and to Bolivia's international credibility as an investment location.'[57] Germans still believed that there was a way out of the rut. First, they believed that the change of leadership of the Bolivian state lithium company (YLB) could bring some progress to the case. With the fall of Morales the change came in the person of Juan

Carlos Zuleta, as a head of YLB. A long-standing critic of Morales' lithium programme, however, he did not intend to revive the deal.[58] Zuleta spent only one month in office,[59] being ousted by another wave of protests.[60] The derailing impact of local organizations over the subsequent lithium projects in Potosí proves their strong position in the larger scheme of things. It is surprising that their leaders and opinions have been ignored for so long by the foreign companies hoping to do lithium business in Bolivia, starting with FMC-LITHCO and ending with ACI Systems. Foreign politicians and executives coming to Bolivia are duped by the facade of a strong-handed government in power. They rarely if ever recognize how fragmented the local political and economic scene really is.

The Germans waited for the presidential elections. The interim president, Jeanine Áñez, had been in power for almost a year, but was citing the coronavirus pandemic as a reason to postpone them, while consolidating her power base. Amid the continuing turmoil in the country, Áñez signed a new law exempting the military from responsibility when using physical force for 'legitimate defence'. Her government has been described by opponents as a mix of devout and militaristic—a picture best exemplified in the idea to sprinkle holy water from military helicopters on coronavirus-infected neighbourhoods.[61] The times were not stable; many wondered for how long Áñez planned to stay in power, and whether the elections would be fair. At Salar de Uyuni, production of potassium chloride from the industrial plant and lithium carbonate from the pilot plant has been badly impacted, falling much below targets.[62]

The Chinese are building a commercial lithium carbonate plant at Salar de Uyuni, and the Chinese joint ventures operating on the other smaller salars have not been cancelled and keep on progressing. The project constructed by Beijing's Maison Engineering on behalf of the Bolivian state-owned YLB (Yacimiento

Litio Boliviano) was supposed to be finished in 2018. The real progress on the facilities is difficult to ascertain. The place is one big construction site. The earlier promises made by YLB, after the first deadline was not met, that the plant would be completed in 2020, were not fulfilled.[63] No wonder, considering the recent turbulent political and economic climate.

In November of 2020, Luis Arce, MAS (Movement for Socialism) candidate in the election, won in a landslide victory, getting over 52 per cent of the vote. The period of the interim presidency, coupled with the healthcare and economic crisis, made people crave the stability that the long-reigning MAS represented. Arce had served as minister of economy and finance under Morales for close to eleven years. Bookish and UK educated, he is seen as a technocrat on the Bolivian political scene. This is despite the fact that he pushed for nationalization during his time in government, believing in Bolivia's unique economic model, where the state looms large. For most of his time as minister, the economy had been doing well, propelled by the commodity boom. As a president, however, he will not have an easy start. During his campaign, he highlighted his interest in making the lithium industry happen. Despite coming from MAS, his programme seemed to propose starting with a clean slate.[64] This may be the most dangerous proposition of all, considering a long history of false starts.

Meanwhile, the Chinese ambassador Liang Yu restated: 'By 2025, China is going to need 800,000 tons of lithium. We are available to help in industrialization regarding metals and chemicals. We are to realize the South American energy and industrial dream of Bolivia'.[65]

Whether Bolivia will start from scratch, or follow the German or the Chinese path, remains to be seen. In a country where lithium is more politically and socially charged than anywhere else in the world, whoever fills the shoes of the country's ultimate, successful lithium developer needs to first deal with the soft aspects of the project. Inclusion of local communities and political stakeholders, infrastructure and educational investments in the region, transparency of the process and its communication to the populace are paramount. The task may prove as difficult on the level of PR as it is on an engineering level.

5

ARE WE REALLY MAKING THE WORLD
A BETTER PLACE?

You have guns, you don't need a salary[1]

Mobutu Sese Seko (President of DR Congo 1965–1997)

Lithium is not a conflict mineral. You will not find a place in the world where revenues from lithium mining support activities of armed groups. Lithium is also not mined artisanally or by children. Due to the geographical locations of the deposits and the complexities of the lithium mining process this is unlikely to change. The same cannot be said of the second most important metal used in batteries—cobalt. Around 60 per cent of the cobalt supplied on the global market comes from the Democratic Republic of the Congo, a Central African country plagued with problems.[2]

Congo typically ranks at the very end of global indexes measuring quality of life, easiness of doing business, literacy or GDP per capita. It makes the top, however, in rankings of the most corrupted countries on earth. The conflict that it experienced between 1998 and 2003 has often been referred to as the Great

African War due to the sheer number of state and non-state actors involved and the casualties suffered. Even though the war has officially ended, it left a number of non-state armed groups still operating (as of 2020), and inflicted a lasting mark on the psyche of Congolese society.

Congo's area is roughly the size of Western Europe, while the number of its population is close to that of Germany. Prolonged conflicts and diseases led the median age of the Congolese to oscillate around eighteen. The country's capital, Kinshasa, is situated in the very west, whereas the middle is covered by impenetrable forests. The east, abutting Rwanda, has for decades been a theatre of guerrilla warfare, sometimes increasing and sometimes decreasing in intensity. The usable roads in DR Congo are scarce and most of the population travels across the country using a network of ferry crossings and air transportation. The lack of infrastructure, together with natural barriers such as jungle, swamps and water bodies, makes it difficult to keep the country united, and limits the reach of the capital over eastern parts. It also leads to regional particularism of the Congolese political scene. The relative weakness of the current president, Félix Tshisekedi, is often explained through a lack of support from any of Congo's twenty-six provinces. His predecessors, as well as most of the heavyweights of Congolese politics, typically enjoyed a strong regional power base.

But we are interested the most in the south-western part of the country, where an amazingly rich copper-cobalt belt—300km long and 30km wide—is situated.[3] The terrain is in what was for decades a Katanga province, until in 2006 President Joseph Kabila, in desperate efforts to hold on to power, further divided the country's provinces, splitting up Katanga into Tanganyika, Haut-Lomami, Lualaba and Haut-Katanga provinces.

Katanga has been a hotbed of geopolitical activity for a long time. Its riches greatly contributed to increasing Belgian wealth through the company Union Minière du Haut-Katanga, which operated on the copper belt from the beginning of the twentieth century up until the 60s. During the Second World War and its aftermath, the US protected the province's rich uranium deposits from falling into Nazi and then Soviet hands. In the 60s, just when the Congo was reaching its long-awaited independence, Katanga's regional particularism was fuelled and militarily supported by Belgian mining interests and then the Belgian state itself, leading to Katangese secession, which lasted for three long years.

Later on, during the bloody Great African War, the province served as a battlefield between diverse Mai-Mai groups and the DR Congo state army. Mai-Mai comes from the Swahili word for water, relating to the guerrilla tradition of sprinkling themselves with holy water, as a supposed means of protection from bullets. Mai-Mai never meant a single militant group, but is an all-encompassing term referring to any kind of community-based guerrilla forces—often based on tribal or village unity in the face of an external threat.

In the twenty-first century, after the conflict faded away, Katanga became a centre of interest for China. From 2002 onwards trade between China and Congo was increasing rapidly. Congolese citizens have been flying to Guangzhou (Canton), returning with Chinese textiles, mobile phones and domestic appliances to resell them on the domestic market with a mark-up of often more than 100 per cent. In turn, China became increasingly reliant on Congo's raw materials for its booming heavy industries and infrastructural investments. Congolese copper, cobalt and wood were in particular demand.

While for copper China has been and still is much more reliant on Latin American resources, and while the Chinese nouveau-riche can make do with other tropical wood varieties for floors in their residences, cobalt is a different story. In 2007, already 85 per cent of cobalt in China came from DR Congo. Today it is 98 per cent.[4] The hard customs data is in striking contrast to the reassurances by some battery, electronics and EV makers that there is no place for DR Congo's cobalt in their supply chains. Simply speaking, there are very few places in the world to go to for cobalt apart from DR Congo. Still if we give those companies the benefit of the doubt and look at South Korea's or Japan's customs data, the only two other countries that play a role of any significance in the production of cathode materials and cathode material precursors for batteries, we will find that they import most cobalt in raw or processed form either from DR Congo or from China, which, as we have already seen, takes it from DR Congo itself.[5]

One out of five tons of cobalt exported from DR Congo comes from artisanal mining.[6] Artisanal mining does not necessarily mean illegal mining. It takes place, however, with the use of the most basic tools, such as shovels, chisels or pickaxes, and typically with very little concern for health and safety. Take your randomly chosen artisanal mining site in Congo's cobalt mining belt. Statistics can be deceptive in such places, but based on data provided by various NGOs, we can safely assume that there are around 100 or more such sites in operation. You will see unprotected men, many of them very young, as sixteen is what constitutes a legal working age in DR Congo, entering 50m-long, very narrow underground tunnels. The heat deep underground is close to unbearable, while the amount of dust inhaled on a daily basis often leads to 'hard metal lung disease', resulting in a range of respiratory problems.[7] Despite all the negative impacts that artisanal mining entails, most NGOs, who are investigating human rights violations at Congo's mining sites, do not advocate the closure of artisanal mining activities as such: too many livelihoods depend on the income, in places where other opportunities to generate a steady income stream are scarce.[8] Different estimates indicate that, just in Katanga province, between 70,000 and 120,000 people work in artisanal mining. And there are of course many others who artisanally mine tin, gold or coltan in other parts of the country. It is not only men that work in mines: women at artisanal sites can be found carrying heavy bags of ore up to 40kg each, later on washing and sorting their content. Children below working age help them in those activities, often to earn the money to pay for their schooling. Even if, in theory, primary education is free in DR Congo, teachers expect parents to contribute with their own funds to their low government salaries.[9] If that does not happen, children are often turned down at the doorstep.

Artisanal mining, as abhorrent as it may seem, allows women to have their own funds, supporting their independence within families and communities.

What NGOs and, at least theoretically, the Congolese government tries to achieve is the formalization of artisanal mining. This should improve working conditions, alleviate health and safety hazards and keep children away from mining sites, while guaranteeing their parents a high-enough income to cover schooling costs. Of course, this is easier said than done.

The long history of exploitation of Congo's copper-cobalt belt that we have briefly described led to a situation where its territory is tightly covered by mining concessions, belonging to a variety of companies and individuals. Indeed, it is very hard to find a patch of land of any value on which to establish a new 'model' artisanal mining operation. If nevertheless the government finds a piece of land that becomes designated for officially recognized artisanal mining operations, miners might still hesitate to come there: such places tend to be less attractive in terms of the amount or quality of the cobalt ore that can be found in comparison to unsupervised sites.

Often the most attractive sites are the ones that belong to large multinational corporations, where principally large-scale mining with the use of industrial excavation machinery takes place. Tailings are what is left over after the mechanical separation of the valuable parts of the ore. Still, some valuable parts are missed by the machines, as the process is happening on a large scale. Artisanal miners come to the piles of tailings, stored on the company's land, often during the night to sift through the leftovers. A mine's territories are sometimes so large that it is difficult to supervise them full-time. Hired guards, when they find the artisanal miners stealing the ore, are more likely to take a cut themselves and let the miners go than to prosecute them. Unofficial arrangements, where a group of miners is paying the security in return for being able to mine on the company's territory, are fairly common. Sometimes it is the company itself that recognizes the value in working with artisanal miners. Mining companies are aware of their machinery inefficiencies and know that at the micro scale there is nothing to match the accuracy of the human eye. Thus they allow artisanal miners to excavate on their sites, retaining a priority right to buy whatever they manage to find.

Scenes such as those from the film *Blood Diamond*, starring Leonardo DiCaprio, where artisanal miners are forced to mine at gunpoint, do not happen for cobalt, today. Based on the current political and military situation in Congo's copper-cobalt belt, this can be stated with some degree of certainty. We cannot entirely preclude, however, that such scenes did not happen in the past, amid the turmoil of the Great African War. Still, cobalt has rarely been classified as a conflict mineral. In fact, many experts argue that treating cobalt as a conflict mineral is harmful to Congo's interests, as well as factually wrong. Conflict minerals are what we call the '3TG'—tin, tantalum, tungsten and gold.

The international legal framework differs from state laws in the sense that it is applicable to countries, not corporate entities or individuals. It is more of a set of guidelines, the incorporation into national law and enforcement of which is left to individual states to pursue.

Such is the nature of the OECD Due Diligence Guidance for Responsible Mineral Supply Chains. Perhaps the only set of regulations that can lead to substantial financial penalties for the companies sourcing conflict minerals is the Dodd-Frank Act. It was introduced in the aftermath of the 2008 financial crisis with the chief aim of being better able to oversee the derivatives market. The section of this act on conflict minerals has been thoughtfully incorporated within it, as its byproduct. As for a kind of 'by the way' provision, section 1502 is very powerful. All the companies listed on the US stock exchange are required to determine if 3TG metals that they use in their supply chains come from DR Congo or a neighbouring country, and, if they do, to carry out due diligence on whether their purchases are funding armed groups there.

Luckily, due to increasing coverage of conflict mineral issues and thus higher social awareness of the problem, the risk for corporations is not only financial but first and foremost reputational. For Apple or BMW, the intangible brand value takes a big chunk of the company's overall valuation. The global consultancy Interbrand has valued the BMW brand alone at $41 billion.[10] The automaker has decided to stay away from Congo's cobalt for now.[11] In order to achieve that, BMW will source cobalt directly from mines in Australia and Morocco. This is definitely a safe strategy and a clever PR tactic, allowing BMW in effect to say: 'we have nothing to do with child labour and armed conflicts in Congo'. But is this really, morally, the right thing to do?

In 2019, Morocco was responsible for only 1.5 per cent of global cobalt production, and its reserves are a mere 0.5 per cent of Congo's reserves by tonnage. Australian cobalt reserves are much higher, estimated in the range of a third of Congo's reserves, but Australian production in the past years has been only two times higher than Morocco's.[12] In Australia cobalt is produced as a byproduct of nickel and copper mining. From a technological, economic and time perspective it will be extremely difficult for Australia to position itself as a major cobalt producer within the next decade. So even if BMW should manage to stay away from Congo's cobalt, this strategy cannot be replicated *en masse* by other companies. There is just not enough cobalt mined outside of Congo to do so.

But for the sake of a thought experiment let's assume that there is. Still, Congo is the only country in the world whose GDP growth rate is directly related to cobalt prices. According to the International Monetary Fund, Congo's extraordinary GDP growth rate from 3.7 per cent to 5.8 per cent in 2017 was driven by the cobalt price spike on global markets. Eighty per cent of Congo's export revenue in general relies on the export of minerals. We are talking here about a country with an extremely small budget in relation to population size. According to estimations based on the World Bank's data, Congo's budget is able to provide only $2 per day per citizen.[13] Is it really a good idea, then, for the world's most powerful companies to stop paying Congo for its cobalt on ethical grounds?

Even if we consider the level of corruption on the ground, which must be eating up a large chunk of cobalt revenues, certain basic needs in respect of infrastructure, healthcare, schooling and security can only be supplied on a comprehensive and sustainable basis through government spending.

Over twenty-one years of the Kabila family's rule did not serve to improve the transparency of business deals in Congo. Kabila père overthrew Mobutu Sese Seko, a colourful yet violent dictator, known for his fondness for leopard skin haute-couture military hats and for institutionalizing corruption in Congo.[14] Joseph Kabila took over in his early thirties, after his father was assassinated by his teenage guard. The event was tragically ironic, considering Laurent Kabila's blind faith in his child soldiers.[15]

MAKING THE WORLD A BETTER PLACE?

Kabila's son was reportedly not ready for the hand-over of power (something that he denies in interviews), but he was a natural choice among his father's powerbrokers, who were reluctant to cede power to any one among themselves. He had solid experience in the army but was lacking political and oratorial skills. Due to the time spent in hiding abroad (for security reasons when his father was fighting the incumbent), his proficiency in French and Lingala, two main languages spoken in Congo, was acutely lacking.

He was derided for a lack of charisma and for playing Nintendo games. In fact, 'Nintendo' turned into his nickname. Nonetheless, his early lack of political seasoning was not an obstacle for Joseph Kabila's hold on power for almost eighteen years and his ability vastly to improve the security situation in the country. Nowadays conflicts are limited and confined to the eastern part of Congo's territory.

It was only during Kabila junior's rule that cobalt became important as a strategic mineral for battery applications. Two of Kabila's connections are interesting from the perspective of our story: the link with the Chinese, and that with a controversial Israeli billionaire, Dan Gertler.

Joseph must have had some sentiments for China from his early years, having spent some time there studying at PLA National Defence University, a top military school in China. He got there as the son of a politically sensitive figure, but later on, China opened up generous scholarship programmes on a mass scale to bright African students of a more humble descent, as a means to securing its influence on the continent.

Kabila junior directly oversaw a huge deal between a consortium of Chinese companies and the Congolese government. The terms of the deal were straightforward—mineral concessions in exchange for infrastructural investments. In 2008, the deal was the one with the largest Chinese footprint in Africa. Out of $9 billion to be invested in the project, $3 billion were allocated for mining investments, and $6 billion for infrastructure.

Some commentators believed that the deal was fair for Congo, since, at least at face value, it limited the potential for corruption. Mineral concessions, the country's national treasure, were not sold for cash that could easily end up in corrupted officials' pockets. They were exchanged for concrete assets such as roads, universities, hospitals and railways.

Others saw it as the steal of the century. The main criticism was that Congo's resources had been hugely undervalued. Awarded concessions were said to contain over 6 million tons of cobalt and over 10 million tons of copper in guaranteed as well as probable reserves. At today's low cobalt prices, just the cobalt alone contained in reserves of such size would be worth around $198 billion.

The wording of the agreement made the deal qualify as a loan in the eyes of institutions such as the IMF and the World Bank, helping Congo under an existing debt relief programme. International organizations were not happy for Congo to take on more debt while they kept busy cancelling the country's previous obligations.

One decade after the deal's conclusion, the assessment by outside observers regarding the state of infrastructural investments is rather negative. The infrastructure already built tends to be of poor quality, and some of the promised projects have not even been started.

Joseph Kabila's administration has not only been selling Congo's mining resources to the Chinese, but has also secured large deals with some of the biggest Western mining and commodities companies, such as Glencore, and Eurasian Natural Resources Corporation.

Dan Gertler, an Israeli businessman, a grandson of Israel's Diamond Exchange co-founder and a personal friend of Joseph Kabila, has been said to be instrumental in these deals.[16]

Gertler started his diamond dealing business fresh after an obligatory stint in the Israeli Defence Forces. His search for new fortune took him to DR Congo in 1997, when the winds of change were blowing.[17] Kabila père had been fighting his war for 'free' Congo since the 60s. He even briefly met Che Guevara, hoping to bring Cuban style revolution to Central Africa. According to Guevara, of all of the people he met during his campaign in Congo, only Kabila had the 'genuine qualities of a mass leader'. It took Laurent Kabila over three decades of mostly armed struggle to secure power in Congo. At the moment when he was in the final stages of the power grab, his son met the twenty-something Gertler.[18] It took Kabila over two years to restore at least partial order in Congo, which was in chaos after his rebellion against Mobutu Sese Seko. He needed money and arms in order to achieve that. The ingenious young Israeli allegedly had a proposal on how to get both. According to a UN report, he proposed assigning a diamond trading monopoly to his company International Diamond Industries, in order to turn the diamonds into cash quickly, as well as access to Israeli military equipment and intelligence.[19] Both were possible, considering the high standing of his family in Israel. The UN argues that this deal has not been a good one for Congo, as local diamond miners and traders preferred to smuggle diamonds to neighbouring countries, where they would fetch higher prices on the free market than under a monopoly, thus stripping DR Congo's budget of revenues from taxes.

Nevertheless, Gertler's relationship with Joseph Kabila developed. Gertler with time grew into the 'go-to' man for mining companies seeking access to DR Congo's resources. He facilitated a number of large oil and mining deals in the country.[20] Kofi Annan's Africa Progress Panel accused Gertler in its report of buying undervalued mining concessions through companies registered in tax havens, using its close links to the president and selling them on at market prices to foreign corporations. Gertler denied that assets were undervalued, arguing that he was buying them at a time of instability, when nobody else was willing to risk investment.[21]

Gertler was also a co-investor, together with Glencore, into the world's largest cobalt mine—Mutanda. Later on, as the relationship between Kabila and Gertler came under more intense scrutiny from the US authorities, Glencore decided to buy out Gertler's stake,[22] valuing it at $922 million.[23] It was a timely move on behalf of the commodity trading giant, as only ten months later, the US sanctioned Gertler based on the Magnitsky Act for alleged involvement in corruption.[24]

Glencore, often referred to as 'the biggest company you never heard of', is a key player on almost every commodity market and has a colourful history that would call for a separate book. From its humble beginnings in Switzerland's sleepy town of Zug, it has grown into an organization with revenues of over $215 billion—a figure higher than New Zealand's annual gross domestic product.[25] This is largely thanks to the talent and hard work of Marc Rich, a controversial figure known for inventing a spot market for crude oil, 'trading with the enemy' (read Iran) and for receiving Bill Clinton's pardon—something that the ex-president later referred to as 'terrible politics'.[26]

The Glencore move was interpreted not only as a way to sever ties with Gertler but also as an attempt to consolidate ownership of mining assets before the EV-driven cobalt boom. Interestingly Glencore, all of a sudden and not long after the expensive buy-out, announced that it was putting the Mutanda mine into care and maintenance for a period of two years, effectively taking 20 per cent of the world's cobalt production off the market. The move might have been calculated as a means to help ramp up cobalt prices after their fall from 2018's peaks, and to preserve the life of the mine for better times, when EV sales would really start to soar.[27]

Lithium mining has its own share of controversies, but they are nowhere near the trouble cobalt causes. The negative impact of lithium mining can be roughly divided into environmental impact, and that on communities living nearby.

As explained in previous chapters, lithium comes from brine or hard rock deposits. Mining from brine is very green as far as CO_2 emissions are concerned. The sun is the source of energy, and it is harnessed to concentrate lithium in the evaporation ponds.

The main concern is the impact of brine extraction on the water in the areas surrounding the salar, as it is used by animals and communities. One thing to state from the outset is that brine itself, despite being liquid, has nothing to do with potable water. Brines by mass are 25 per cent salt and 75 per cent water—which seems still to be a lot of water, until we consider that sea water is only 3 per cent salt by mass, and the water that we drink contains less than 0.1 per cent salt.[28]

So, it is not such a great waste to lose this extremely salty water, even in an arid environment. What is more worrying is how pumping off the brine affects the potable water aquifers near the salar. Movement of water is studied by hydrologists, who built complex hydrological models. The problem is that at the moment we lack the data to know with certainty what the impact is. Companies who mine lithium collect data, but this data is not available in the public domain. Even if it was, we would never know if it was not biased. We also would lack comparable data set as a benchmark.

One of the worst-case scenarios is that potable water from the outskirts of a salar might be sucked in when the brine is extracted. Such a scenario seems plausible from a lay person's perspective, but we do not know whether this is really what happens. In a court of law a person is innocent unless proven guilty. When companies deal with regulators, the opposite might be true—a Chilean producer's application for lithium production expansion was rejected in an environmental court, just because the company was not able to prove that potable water aquifers are not affected.[29]

With all the sun and wind on the salars, adjacent water reservoirs are evaporating anyway, with or without lithium mining. It further complicates the matter of determining the miners' impact.

Brine-based extraction is in any case a lesser evil if we do not want to give up on batteries. Hard rock extraction certainly causes more harm, as the spodumene concentrate derived from rocks needs to be roasted at a temperature of 1,050°C. Then it needs to be cooled, mixed with highly toxic sulphuric acid and heated again. Heating, re-heating and drying processes are CO_2 emissions intensive.[30]

Fresh water is typically used in the leaching, flotation and washing stages of hard rock-based lithium production, and according to some of the experts, the total use of water in spodumene mining is higher than in the brine extraction route.

Again, as in the case of the salars' environmental impacts, we need more independent studies to determine real CO_2 production levels, across the battery supply chain. Especially since the industry is changing so fast. One of the most authoritative studies on the CO_2 footprint of the battery industry often cited by the media is Argonne National Laboratory's assessment. It was published for the first time in 2012, when the majority of lithium still came from brine-based deposits, where extraction is a process driven by solar energy.[31] In 2020, a much larger share of the lithium ending up in batteries came from hard rock. The Argonne study determined that 2.5t of CO_2 is produced per each ton of lithium compound. Nowadays, the CO_2 amount per ton could easily be more than seven times higher.[32]

Even in the case of 'green' production from the brine, the CO_2 amount can be doubled, as the battery world transitions from using lithium carbonate to lithium hydroxide (which requires more processing) in high-performance EV batteries.

The impact of the reagents used in the brine-based production process also needs to be considered. Evaporation-based processes rely on large amounts of lime and soda ash. Soda ash is considered non-toxic—in fact it is used in areas where acid spills occur in order to neutralize them. Still it needs to be handled with care, as it is dangerous to breathe in soda ash dust. Lime, used to remove magnesium impurity, which is detrimental to battery-quality lithium, has caused concerns among Bolivian activists, who are afraid of accumulation of mountains of residual sludge, spoiling a pristine salar landscape.[33]

Lithium extraction waste is already becoming a problem in Western Australia, home to the largest lithium spodumene mines in the world. One problem with mining is that it is always an invasive process. You are tearing the earth to get what you need, and, no matter whether it is lithium or coal, you leave a scarred landscape behind—full of pits and piles of tailings that need to be stored somewhere.

Dardanup, in Western Australia, is a quaint little town, situated in a fertile, wine-growing region. The landscape is truly picturesque. Since the boom for lithium mining began, the area just outside the town became the end destination for the tailings of the world's largest lithium spodumene mine. The plan is to dump there 600,000 tons per year of leftovers from ore processing. Needless to say, the local community is not happy.[34]

When we move up the value chain, to the stage where battery materials are turned into cathode materials, the heart of lithium-ion batteries, the problem with CO_2 emissions continues. Cathode materials precursors needs to be calcinated (i.e. heated at very high temperatures), numerous times, in large kilns that are never put out (as on and off ignition would not be economic).[35]

Both spodumene concentrate processing and cathode materials production are dominated by China, and the pollution generated becomes effectively a Chinese problem. China is the world's largest CO_2 emitter, responsible for around 30 per cent of the global share. When we look at the carbon emissions problem, we should consider it from the perspective of scale and intensity. For instance, the amounts of lithium hydroxide or cathodes materials produced in China are nowhere near the amount of cement that is fuelling the country's construction boom.

The amounts of lithium hydroxide produced in China in 2019 were estimated at 76,000 tons.[36] If we assume even 15 tons of CO_2 per ton of lithium hydroxide, this adds up to 1.1 million tons of CO_2 released into the atmosphere. At the same time, China produced 2,250 million tons of cement.[37] The production of a ton of cement is seven times less polluting in terms of CO_2 emissions than the production of lithium hydroxide (though tell that to EV makers who promote their vehicles with zero-emission slogans). But still hydroxide is responsible for less than 0.01 per cent of the CO_2 emissions that cement generates in China.

We can do more to keep the battery supply chain from mine to battery pack more sustainable. In the end, though, EVs will be only as green as the electricity that powers them. The Chinese electrical grid runs primarily on coal, which represents close to 65 per cent of the electricity generation mix. Renewables take 28 per cent, nuclear 5 per cent and gas a mere 3 per cent, according to the newest Q1 2020 International Energy Agency data.[38] For the sake of comparison, in 2020, Germany generated 24 per cent of its electricity from coal.[39] Coal is particularly nasty to burn as it produces up to twice as much CO_2 as other fossil fuels.

In the past years, China has been responsible for almost half of the renewable energy capacity growth globally. However, considering that China's energy appetite is stimulated not only by the world's largest population but, even more importantly, by its massive industrial base, it will be an uphill struggle to reach levels close to a much greener but still very industrialized economy such as, for example, Germany. It can even be argued that, considering coal-based power generation in China together with the battery production process, EVs are on a par with if not more polluting than conventional cars in terms of CO_2 per km levels.[40]

But again, this is a changing landscape. In 2012 in the US, less than half of the population lived in regions where the average EV produced fewer emissions than a conventionally powered car with a good fuel economy rating. In 2020, nearly everybody lived in such an area. What kind of electric car you drive and where in the world, determines to a great degree how much less you emit in comparison to using a gasoline car. Sometimes, the difference you make in terms of greenhouse gas emissions might be disillusioningly small. But if you drive a new Tesla model in California, your carbon footprint may be as much as 60 per cent lower than that of the most efficient gasoline car.[41]

It is not only the emerging economies, though, which are facing a problem with coal-derived energy polluting the environment. The cradle of battery technology, Japan, otherwise a high-tech nation boasting achievements in semiconductors, robotics, specialty chemistry and advanced materials, runs on carbon. Almost a third of Japan's generated electricity comes from coal.[42] Japan's relation with black gold is convoluted. Oil was its main source of energy up until the 70s, when members of OPEC effectuated an oil embargo on countries supporting Israel in the Yom Kippur War, with Japan among them. Since then, coal has been a fundamental source of electricity for Japan, supported by nuclear power generation, which completely lost its appeal after the Fukushima disaster. The relatively modest advances on the renewables front can be excused because of the country's geology, with its rugged terrain not fit for solar panels, and the very deep seabed close to its shore, which makes windfarms a challenge. It seems that the structure of Japan's energy mix will not change much any time soon. Until 2025, another twenty-two coal burning power plants are in the pipeline.[43]

It would be a mistake, however, to focus solely on car engines as a source of greenhouse gases. CO_2 even in relatively high quantities has no real impact on human health. The same cannot be said about particulate matter (PM) and nitrogen dioxide, which are also coming out of exhaust pipes. Modern engines are a source of the very harmful PM 2.5 particles. They are only 2.5 micrometres or less in width, which makes it easy for them to get through the body's natural defences situated in the nose and lungs, which are efficient at filtering out larger particles. While climate change is by definition global, and we need to look at an aggregated level of emissions to solve it, from a health perspective, it actually matters a lot whether there are more or fewer EVs driven in your neighborhood. Nitrogen dioxide pollution stays in the air for less than a day, and it does not travel far. At the same time, even the smallest particulate matter radiates only metres from its source. Due to their size they easily reach our bloodstream, carrying metals, coated with petrol or diesel. They are carcinogenic and are associated with causing or aggravating asthma. Existing studies point to reduced lung capacity in children growing up in areas with nitrogen dioxide and PM levels much above the norm. So even if you cannot be entirely sure that by buying an EV you will contribute to stopping global warming, you can be certain that you are improving the quality of air you and your children breathe.

There are metals other than cobalt and lithium that are necessary for the functioning of some of the more popular battery types. Graphite is key for the production of the anode (the battery's negative electrode), while nickel plays a very important role in the cathode, the positive electrode in NCM batteries. Elon Musk in fact pointed out that the name 'lithium-ion battery' was a misnomer, since, as he argued, there is much more nickel and graphite in the battery powering Teslas than lithium.[44] The markets for nickel and graphite, however, are driven mainly by non-battery applications, especially steel production. Nickel improves steel's strength, toughness and resistance to corrosion, while graphite is used as the protective agent in steel ingots, and also lines metallurgical furnaces. Nickel's use in batteries accounts for around 3 per cent of its total demand, equalling almost 2.5 million tons.[45] Therefore, for most of the nickel and graphite miners, the battery industry is a new and exciting source of demand, but in the larger scheme of things, it is not very important. What further complicates the matter is that not everybody in nickel or in graphite has a product that is suitable for the battery industry, because batteries do not need nickel per se, but rather nickel sulphate, a processed nickel derivative that until very recently has been an obscure niche product.

It is really not a contest, but the environmental impact of the graphite and nickel industries seem to beat that of cobalt and lithium. Graphite, pretty much the same substance that you find in pencils, is a form of carbon, and everything carbon-related is by definition dirty. Unless, that is, we speak of diamonds, which graphite can theoretically convert to if you apply high-enough pressure and temperature. Graphite can be mined and artificially produced. For battery applications, both routes are viable, though mined graphite is much cheaper and thus often constitutes the first choice for anode active materials producers. No matter which route you take, graphite typically occurs in the form of flakes, and, to be fit for anodes, needs to be further shaped and purified. As usual, mining is easier than the later chemical processing step. China is by far the biggest miner and processor of graphite.[46] Mozambique entered the stage fairly recently as an important exporter of graphite, with a mine developed there clearly with batteries in mind. Germans invented the chemical processing step, but they did not care to patent it, and so Chinese players stepped in and practically took over the entire spherical graphite production. Developed countries did not mind so much, because the process is polluting and margins are low. China was able to improve the margins by applying cheap labour where machines are typically used. They also did not reject the use of hydrofluoric acid in large volumes, which is an integral part of the process. When you enter the Chinese village of Mashan, one of the key locations on China's graphite industry map, some people are clearly proud of their choices. Billboards promote graphite as a green, high-tech material, one that fosters change.

The change is for better, but not in Mashan. The district boasts the type of landscape that one might typically associate with China. It looks as though taken from ink-painted scrolls—rice fields with singular, rounded green hills scattered among them. Mashan had been making a living out of agriculture and tourism before the graphite processing facilities moved in. It took some time before locals started to notice the visible signs of graphite production. The air became so dense with soot that reflected light made the evenings sparkle. Crops started to suffer, and water from many of the traditionally used wells adjacent to the houses became unfit for drinking.[47]

'Heaven is high, and the emperor is far away,' as the centuries-old Chinese proverb goes. Places such as Mashan, where raw materials are mined or processed, are far away from the centres of power. They are far away, too, from where media, NGOs and authorities independent of local coteries are based.

The reporters from the *Washington Post* who first brought Mashan's graphite issues to global attention described the atmosphere of intimidation in the area. As they went about the town asking questions, they had a feeling of being followed. Locals considered going on record or even talking to the press a risky endeavour.[48] One would think that real danger would be highly unlikely, but the high-profile case of an analyst investigating a Chinese silver mining company over short selling serves as a cautionary tale. Short sellers borrow shares of companies that they believe are over-valued and sell them on immediately at high prices. Then they wait for other market participants to come to the same conclusion in order to buy the shares back at the lower price, profiting from the difference. Sometimes they help the process of bringing the price down along by publishing reports stating their arguments against the targeted company. Companies hailing from China, seeking capital on Western stock exchanges by registering and listing in the West, quickly became infamous for posting—delicately speaking—overly optimistic annual or quarterly reports.

The Canadian stock analyst Kun Huang went to great lengths to verify information published in Silvercorp's reports. He travelled to a small town in Henan province, where he filmed trucks leaving the mine's premises in order to verify the volume of shipped ore. He also collected the material that fell off the trucks in order to independently check its quality. Then he made his findings questioning the company's performance public. When the company alerted local authorities of his activities, he was arrested, forbidden from leaving China and then sentenced to two years in prison. He had committed 'the crime of impairing business credibility and product reputation', according to the information *New York Times* received.[49] This is an outrageous charge for an activity that in the end leads to more informed and transparent capital markets.

The last element key to the functioning of a high-power EV battery is nickel, which does its own share of environmental damage, also outside of China. To illustrate the gravity of its impact, one might consider the Russian arctic town of Norilsk, which was home to both a nickel mine and refining facilities. It was ranked as the most polluted town in Russia and indeed in the world, until the smelting facility closed in 2016.[50] In the meantime, the operation had been spewing over a million tons of sulphur dioxide every year. The pollution was so intense that at times, the inhabitants were able to sense a sulphuric taste in their mouths, just by breathing Norilsk's air.

At present, Norilsk Nickel invests heavily in becoming more sustainable, partly based on environmental pressures from the Russian administration and partly as a requirement to align with investors' expectations. Nowadays, large investment funds more often need to follow a mandate to invest in securities that score high on environmental merits. Still, accidents may happen. In 2016, the Daldykan river close to Norilsk turned a biblical-looking red.[51] Heavy rains led to spillage over the mine's filtration dam.

The world's two biggest miners of nickel—Indonesia and the Philippines—take a distinct approach towards their nickel industries. Indonesia tries to position itself as a battery and EV production hub, hoping one day to provide the rich markets of Australia, South Korea and Japan. Until recently, Indonesia has been mostly mining nickel and shipping it out to China for processing. Now it wants to carry out the value-added but environmentally costly processing at home.[52]

The authoritarian president of the Philippines, Rodrigo Duterte, has on the other hand already threatened to 'tax miners to death' if they do not clean up their act.[53] Under his stay in office, operations of twenty-eight of the country's forty-one mining companies have been shut down following environmental audits. The mines accounted for half of the Philippines' nickel output and reportedly were shut for leaving rivers and fields polluted with red, nickel bearing laterite.[54]

Both countries are famous for their biodiversity, pristine tropical forests and coral reefs, home to clownfish—portrayed as the main character in the popular film *Finding Nemo*. In that context, perhaps the most shocking tactic is the application of deep-sea tailing disposal—a practice where companies dump mining or processing waste into the sea. Indonesia already has companies who do that, and the nickel miners who have smelting facilities in the pipeline apparently want to join in. In January 2020, two nickel producers presented plans to use the deep-sea tailing disposal to Indonesia's Coordinating Ministry for Maritime and Investment Affairs. We are talking potentially about dumping millions of tons of waste into the waters of the Coral Triangle through a network of pipes discharging at depths of 150–250 metres.[55]

But not everybody in Indonesia is as happy for the battery economy as the administration. Residents of Kabaena Island, aware of the fact that they sit on nickel—the island's red soil serves as a tell-tale sign—registered their ancestors' land as a tourist village. Fear of miners moving in incentivize them to maintain their indigenous character, for instance through the organization of an annual festival, even if the level of interest from tourists is rather mediocre.[56]

When we speak of mining, it is close to impossible to do no harm. It is by definition an impactful industry, because only through sheer force can the natural treasures of the earth be harnessed to build our civilization. As a society, we are extremely reliant on the mining and chemical processing industry. Almost all of the material things we have and use on a daily basis either rely on the elements coming from the ground, or have gone through some sort of chemical processing. Batteries, due to their level of complexity, are no different. What is important are information and oversight. We need the objective data to make informed decisions, and in most cases from an environmental perspective, to choose the lesser evil. We also need an independent oversight with instruments, if necessary, to enforce their decisions. At the moment, this is lacking. Society at large remains widely uninformed and prone to green slogans with little substance. At the other end of the scale are environmental activists, who block commercial activities and thus progress without presenting alternatives. The current system is mostly based on these counter-balancing forces. Rampant commercial interests often clash with fundamental environmentalists, resulting in something that, even if does not constitute an equilibrium, at least keeps tabs on the industry.

Notably there is also a global community of geeks, mavericks and entrepreneurs who seek a middle way to limit the environmental impact. They may not resign from chemical processing, but at least they want to cut off the mining part. Transforming natural resource mining into urban mining is a topic that we will tackle in the next chapter.

6

URBAN MINING

Jane Jacobs née Butzner was truly a renaissance woman. She never finished formal education, dabbling in fields as various as law, zoology and economics. She saved SoHo from becoming a part of the Lower Manhattan Expressway and considered 'suburbs' to be parasitic.

She married a guy who designed warplanes and published *The Death and Life of Great American Cities* (1961)—a seminal work as far as urban planning is concerned, coining terms such as 'social capital' and 'urban mining'. She professed at the end of the 60s that cities might 'become huge, rich and diverse mines of raw materials. These mines will differ from any now to be found because they will become richer the more and the longer they are exploited'.[1]

In fact, if we compare reserves of metals playing important roles in our civilization such as gold, silver, lead or zinc with their amounts contained in various material objects in use or in rubbish, we will find out that there is more of the latter available.

What is more, metals contained in stuff we use occur in much higher concentrations than they do below ground.

Battery materials are no exception here. If you want to produce one ton of lithium, you need to process 250 tons of spodumene ore or 750 tons of brine.[2] In this process you will create a lot of leftovers which need to be dealt with. If you take a look at spent lithium-ion batteries, you need only 28 tons of them to get your ton of lithium.[3] This is a remarkable efficiency on a theoretical level.

The abundance of metals above ground and their concentration did not escape the nations who are used to dealing with scarce resources. To an extent, scarcity defines Japan. Scarcity of land and housing is symbolized by cramped Tokyo apartments, which have almost become an element of pop-culture. In the 1980s, when Japanese real estate prices were at their peak, somebody estimated that the Imperial Palace might be considered to be worth more than all the combined real estate in California.

Fewer people know that Japan also experiences scarcity of an even more basic element than land—potable water. The latter seems counterintuitive, as Japan is an island and it seems to always rain in Japanese horror movies. But Japan's mountainous terrain means that rivers are short. Meanwhile precipitation is heavily concentrated in the typhoon season, when rainwater quickly flows to the sea. In fact, Japan is so dry that per capita water availability in the greater Tokyo area is comparable to levels found in North Africa and the Middle East.

The country is also poor in natural resources—it frequently features as one of the top three largest importers of oil, coal and LNG, despite its unassuming size in comparison to China and India who typically find their place on the podium as the biggest energy importers.

It is not a surprise that the idea of urban mining took a hold in Japan, to the degree that the country claimed it as its own. Japanese sources say that it is Professor Hideo Nanjyo of the Research Institute of Mineral Dressing and Metallurgy at

Tohoku University who is the father of urban mining, having come up with the term in the 1980s.

The coronavirus pandemic has only exacerbated the trend. Japan's 127-million-strong economy, with a robust industrial base, felt a sense of isolation when metals shipments were delayed and business travel cut. At the same time, the Japanese fascination with all things electronics—from robot dogs to game consoles—led to a situation where heaps of electronic waste are discarded every year. According to estimates, Japan has more gold in its electronic waste than South Africa in its reserves.

The country's electronic waste recycling industry has grown from processing 100,000 tons in 2005 to over 300,000 tons now.[4] The government is supporting the industry to grow into a profitable segment of the economy, which is more than just an environmental initiative. Harvesting gold and silver from old circuit boards or copper from urban waste already pays handsomely.

Olympic Games are always used as a pretense to showcase the host country's strength. For the Tokyo Olympics, Japan plans to further strengthen its image as a high-tech and environmentally sustainable nation. Toyota plans to present to the world the first car powered by solid-state batteries.[5] This technology will be offering more safety and energy density, and it will also consume more lithium. Self-driving cars are supposed to carry visitors around on a scale not witnessed yet in Japan. Meanwhile the medals given to athletes are to be made only from recycled gold, silver and copper.

Despite the successes with other metals, recycling of lithium or cobalt still remains in a nascent phase in Japan. The country began its journey with the electrification of vehicles and set a goal for every car produced by a Japanese automaker to be electrified by 2050. But somehow, it got stuck in the middle, relying on hybrid propulsion systems, with electric vehicles powered purely by batteries taking only a small percentage of the market share. The Japanese seem to be more reluctant regarding the adoption of fully electric vehicles than folks in China, the EU or the US.

This is reminiscent of the situation with cashless payments. Japan was one of the first countries where you could pay without using cash, but now the country's share of cashless payments lags behind Britain, South Korea and Poland.

Japan's leading automakers such as Honda and Toyota have long been reluctant to turn to lithium-ion batteries. The extremely popular Toyota Prius still used a nickel metal hydride battery in some of its models produced in 2019.[6] This is the type of battery that suffered a dramatic loss in its market share in EVs within the last decade.

Low volumes of fully battery powered vehicles and a long-standing reliance on the nickel metal hydride battery do not make the Japanese market attractive for lithium-ion battery recyclers looking to harvest cobalt and lithium. Nevertheless, Japan is not short of companies eager to do it anyway, more focused on the future than on the present. JX Nippon Mining & Metals Corp. was one of the pioneers in the field. The company, whose core interest was in conventional copper mining and smelting, pivoted towards electronic materials used from semiconductors to batteries. As early as 2010 they put their pilot Tsuruga plant into operation, being able to recover cobalt, lithium, nickel and manganese from lithium-ion batteries scrap.[7]

The process starts with the manual removal of casings and connectors from discarded batteries. Then the batteries are put in a furnace to evaporate the liquid electrolyte. Leftovers are crushed, resulting in a fine and coarse black dust. This dust contains cathode materials with valuable metals inside, and it can easily be separated based on particle size difference. In the next step, metals are extracted from the fine dust using a leaching process.[8] Making a cup of tea or coffee is an everyday example of leaching in action. It is the process of extracting the substance from a solid material with the means of a liquid. As hot water extracts coffee from ground coffee beans, so the solvents extract nickel, cobalt and lithium from fine ground battery dust containing cathode materials.

Since 2010, the company targeted lithium-ion battery waste coming from mobile phones and laptops. Reportedly it was able to recover around 100 tons of metallic cobalt and nickel per year during that time.[9]

The existence of battery recyclers depends on the availability of lithium-ion battery scrap and the prices of the metals contained in it. It is hard to say with full certainty how long an EV battery will serve you. For older EV models, it was between five and eight years on average; in newer ones, automakers promise as much as fifteen years of trouble-free service. A lot depends on intensity of use—expressed as the number of charges or miles driven—as well as climate, as batteries age quicker in extreme temperatures. This makes it likely for an EV battery pack to last less long in Reykjavik or in Dubai than in London. Experts who discuss battery durability think in cycles. Cycling is a bit of a tricky concept. You achieve one charging cycle when you fully discharge the capacity of your battery. The capacity of a lithium ion-battery, however, is not a constant. It diminishes with usage. EVs and some premium electronic devices need at least 75–80 per cent of the initial battery capacity to function well. When the capacity falls below this threshold, batteries need to be replaced. Back to cycles—if you use 70 per cent of capacity today, recharge it to full capacity overnight and then use up 30 per cent until dinner the next day, you achieve one full cycle. The fact that you charged up a little in between is ignored in calculating cycles. The cycle life of batteries is often quoted as their vital parameter, expressed in the number of full cycles.

China will be the first country where used EV battery packs will result in massive amounts of battery waste. 2014 was the first year when EVs in China sold in tens of thousands of units, and in 2018 sales exceeded one million for the first time.[10] In 2019, more than 2 million EVs were sold globally.[11]

If one adds to these 851.2 million smartphone users, plus all the laptops and tablets in the country, all of them running on lithium-ion batteries, then that adds up to some serious urban mine stock.[12] The first wave of a lithium-ion battery waste tsunami is set to hit China before 2025, when batteries coming from the first EVs sold on a mass scale will come of age.

The stakes are high—even if batteries stored in landfill are not considered to pose a serious environmental risk, it is difficult to prevent them from causing massive fires. Besides, they may still leak toxic electrolytes, leading to the contamination of soil and ground water. But China seems to be ready to embrace and transcend battery waste into something if not positive then at least environmentally acceptable. This is from a legal, economic and technological perspective.

If there is one state agency in China that deserves to be watched in the context of new technologies, it is MIIT—the Ministry of Industry and Information Technology. Established in 2008, roughly when China started its technological ascent, it became one of the leading actors in the technological rivalry between China and the US. Its role is not only regulatory, as it has been responsible for the creation of the 'Made in China 2025' plan—a nationwide strategic plan to move away from being a 'factory of the world' to becoming a technological powerhouse. The agency is responsible for everything high-tech in China, starting from AI, through 5G and ending on batteries and electric cars, including their post-mortem utilization.

The agency is led by Xiao Yaqing. He made his name as a leader of China's main market regulatory authority, the watchdog for state-owned companies and, last but not least, as a top executive of Chinalco, China's leading aluminium producer, where he played an instrumental role in steering the company into profitability.[13]

As early as in 2018, MIIT issued the interim measures which placed the responsibility for recovering used EV batteries on automakers. The regulations also made battery producers responsible for encoding EV batteries according to national standards, and for sharing the data with authorities, so as to ensure a fully traceable system for the whole life cycle of each battery. As one of the EU officials busy with recycling initiatives remarked, in the EU each egg that you eat for breakfast can be traced back to the farm. If we have these levels of transparency with so cheap and short-lived a product as an egg, the same should be possible for lithium-ion batteries. China very quickly turned the 'should' into a 'must', and is already doing it.

Putting serial numbers on batteries represents another advantage related to their dismantling process. The more automated and streamlined the recycling process is, the cheaper it gets. And the cheaper it gets, the more universal it becomes. One of the biggest obstacles to bringing down the cost of recycling comes from the fact that lithium-ion batteries come in all shapes and sizes. Even if we put aside electronic devices, where the battery form factor needs to be adjusted to fit each of the myriads of laptops, tablets and smartwatches models, we would still be stuck with many types of EV batteries. One might think that an object as large as a car would make EV battery standardization easy, at least in regard to shape and size. So far, however, this has not been the case. Automakers justify using lithium-ion batteries in different shapes as best suited their individual models and platforms, as they seek an edge over the competition in this still relatively young industry. Standardization across products is a characteristic that is shared mostly on mature markets.

EV battery packs are made of battery modules. In fact, a battery pack is just a collection of battery modules, made of single lithium-ion cells. Those single cells can be cylindrical, prismatic or pouch shaped.[14] Cylindrical lithium-ion cells resemble the commonly used AA alkaline batteries that power less energy hungry devices such as analogue wall clocks or remote controls. Even those cylindrical cells come in different sizes, and different sizes might be used even by the same producer. Tesla touted its new 4680 'fat' cell during the widely publicized Battery Day, providing superior electrochemical performance at least partly due to its wider 46mm diameter and 80mm length.[15] This proves that in a cutting-edge industry such as batteries, tinkering even with small details on the engineering side may provide you with more competitive advantage. Prismatic cells, also enjoying popularity among automakers, are remininscent of rectangular Lego blocks. Since prismatic cells are larger than cylindrical cells, you need fewer of them to assemble the battery module and thus the battery pack. The third type of EV battery is a pouch battery— imagine opening a paper box with a small tablet neatly packed in a silver pouch. Or think of an airtight pharmaceutical pouch packaging for diet supplements. This is close to how an EV pouch cell looks.

Unsurprisingly, one would need a really smart robot to segregate and dismantle such different looking battery waste. That kind of a robot does not exist yet. The field of robotics that deals with object recognition is called machine vision. A machine vision system uses a camera to view an object and computer algorithms to process and interpret the image, before instructing robotic arms to act upon that data. Algorithms that propel advances of machine vision at the moment are based mostly on machine learning—the ability of the computer to improve automatically through the experience. In a famous coding competition, the machine learning algorithms were fed 25,000 labelled pictures of cats and dogs, in order to learn the difference between the two. Then 12,500 unlabelled pictures were fed back into the software, with a requirement to sort them into cats and dogs. The best algorithm during the competition was able to do this with 98.9 per cent accuracy.[16] Machine learning and machine vision have advanced since. Yet in the case of batteries, we need to deal with 3D objects, which are much more difficult to classify than flat images.

In 2019, Apple announced the expansion of its recycling programme.[17] The company made substantial progress from the time when it launched its pioneering automated discarded iPhone disassembly line that could dismantle only iPhone 6 models. The intact discarded iPhones 6 had to be accurately placed on the disassembly line which used pre-programmed moves of twenty-nine robots in twenty-one steps to dismantle them.[18] Currently the robot, which goes by the name of 'Daisy', can dismantle fifteen different iPhone models at the rate of 200 per hour.[19]

We should keep in mind, though, that this is a robot made by the smartphone producer, which knows its devices inside-out, and which is programmed to recycle only its own products. This is another reason why the Chinese legislation which puts the responsibility to recycle on the end-sellers of products containing lithium-ion batteries makes sense. The tagging of batteries by standardized serial numbers is also very helpful, as it will allow robots of the future to rely on information contained within the batteries, instead of just on their imperfect ability to 'see'.

China's battery recycling landscape is something of an enigma to Western market participants and observers. Start-ups and more mature companies from outside China that work on new recycling projects, typically with small processing capacities, act as though they are changing the world and building the future—when the future of recycling has already arrived, in China.

Chinese battery recyclers are roughly divided into large companies, the size of GEM or Huayou Cobalt, and small, family-run, 'mom-and-pop' companies. The latter usually do not handle EV batteries, seeing them as too complex and dangerous to recycle. EV batteries are high-voltage, high-capacity batteries, and if handled improperly there is a risk of life-threatening electrocution. Generally, the skill-set needed to provide even the most rudimentary repair and maintenance service around EV batteries is severely lacking on the global market.[20] This situation needs to change quickly, especially as tending EV batteries without adequate training can result in fatal accidents.

Chinese mom-and-pop operations most commonly rely on physical method and deal only with LCO batteries that power electronic devices. In such operations, valuable cathode materials are manually separated from the current collector.

In the other category, there are companies such as GEM. GEM is a waste disposal behemoth. The company processes around 4 million tons of waste in China.[21] The battery business is just one of their businesses for them, as GEM recycles everything from plastic to motherboards.

It retrieves from waste minor metals such as tungsten, which has been known as the 'perfect metal for bullets', and rare earths, with a number of strategic high-tech applications. GEM's collection network covers eleven provinces across a distance of 3,000km. With its investment in South Africa and Indonesia, it even played a part in the Belt and Road Initiative, China's trademark global infrastructure development strategy that aims to connect Asia with Africa and Europe via land and maritime networks while projecting China's cultural, political and economic clout.[22] The recycling industry is not very flashy, so companies such as GEM tend to stay in the shadows. Nonetheless, publicly available sources state that the company has the capacity to process around 300,000 tons of waste batteries.[23]

Considering that the total of the discarded lithium-ion batteries in China in 2020 adds up to something around 500,000 tons, this is a lot of capacity for one company.[24] For the sake of comparison, the biggest European EV battery recycler has a lithium-ion battery waste capacity below 10,000 tons.[25] Added to this, the European companies that have lithium-ion battery recycling capacities, or are openly claiming to invest in them, can be counted on the fingers of two hands. The situation in North America is not much different. The industry gossips that meanwhile Chinese players are looking at the European and US battery market with whetted appetites.

To get the precious battery waste, the Chinese have two options— either to invest in new plants on the ground in the West, or to buy up the waste and transport it to China. The latter may likely be a more viable option, considering that there is plenty of spare capacity to process it already built. But the transportation of lithium-ion batteries is far from an easy matter. If you attend recycling conferences, once in a while you will stumble on an individual sitting on a potential gold mine of tons and tons of lithium-ion batteries. Such an individual will usually be scratching his head on how to get them to a place where the valuable metals can be retrieved, while not breaking the law or accidentally sinking the vessel.

Most serious shipping lines, including the largest ones, decline to transport lithium-ion batteries. As a willing party, this leaves you with private ship owners. Outsiders would find it intriguing how large a percentage of the world's bulk vessel fleet is in the hands of private ship owners or shipping families. Among those, often with Greek or Norwegian nationality, you may find the daredevils willing to dabble in business that might one day grow big. But even the most risk prone ship owners will not take batteries on board that are heavily damaged. The risk of fire on the vessel would just be too high. In January 2020, a fire broke out on a COSCO Pacific vessel while in the middle of the Arabian Sea. The cargo had been loaded in Port of Nansha, China, with India as the final destination. COSCO have stated that the battery cargo had not been properly declared—figuring as 'spare parts and accessories'.[26]

There were no injuries, but the vessel had to berth in the nearest port. Lithium-ion battery fires are difficult to put out as burning batteries produce their own oxygen, turning them into a self-perpetuating threat. Misdeclaration of cargo to avoid scrutiny, even if at risk of fines, is a common problem in shipping.

There are more dangerous cargos, though, than lithium-ion battery waste that travels the high seas. Few people know that from 1969 to 1990 there were more than 160 shipments of used nuclear reactor fuel from Japan to Europe, which UK-and France-based facilities undertook to reprocess. What is more, the waste left over from reprocessing as well as the resulting fresh nuclear fuel were shipped back to Japan. If humanity endeavours to ship back and forth highly dangerous nuclear waste on one of the longest sea-routes, because it is necessary from an environmental perspective or because it simply pays to do that, transporting something as mundane as battery waste should not be a problem when a real need arises.

If you happen to wonder what type of packaging is used to transport the nuclear waste, we are talking about robust structures in the shape of flasks, weighing around 100 tons per unit. They are covered with 25cm-thick layers of forged steel. A single such structure holds a set of stainless-steel canisters which contain in turn a layer of glass waste around the radioactive material to help isolate the radioactivity. But the ratio of packaging weight to actual cargo weight is around ten to one. Surely, there must be more cost-efficient ways to move battery waste. There are start-ups on the market as we speak, offering special containers to move undamaged lithium-ion battery waste with fire retardant granulate poured in between the cells to isolate them. Nuclear waste shipments require special vessels. If EVs really scale up and displace conventionally powered cars from the market, battery transportation and battery waste transportation could become an attractive niche to accommodate for shipping companies willing to step in and innovate.

Regulatory initiative and oversight are also needed to make sure that battery waste does not end up in the wrong places. Developing countries are flooded with waste coming from the world's more affluent regions. For municipal waste management companies, it is cheaper to ship all kinds of waste over large distances than to utilize it at home, where they would have to adhere to more stringent local standards. This represents an out of sight, out of mind type of mentality. Unscrupulous companies in developing countries often make pledges to process the waste according to innovative technologies. When the waste ends up in the country, before it gets processed in facilities which often have not yet been built, it often burns in an 'accidental' fire, emitting toxins into the air and harming the local populace.

Today, the shipment of lithium-ion battery waste involves very complex paperwork, which presents a heavy regulatory burden even for the larger companies. It falls under both dangerous goods and waste shipments frameworks, and takes over half a year to sort out, even for a single shipment. Paradoxically, the complexity of the legislation often increases the number of available loopholes.

But is it likely that anybody would deliberately burn or landfill EV batteries, when they contain so much valuable metal? Wouldn't throwing them away or burning be equivalent to throwing away or burning barrels of oil?

As market prices of metals fluctuate, so does the value of their content in an EV battery. Studies from 2019 hinted at a value of metals as contained in battery waste of between $5 and $8 per kg.[27]

A barrel of oil weighs on average 136kg. The average oil price in 2019 was at $64 dollars per barrel, which equals only $0.47 per kg. From this perspective, discarded lithium-ion batteries could be more valuable than barrels of oil, if the cost of recycling is low enough.

GEM is already selling substantial amounts of cathode materials precursors—which could be described as 'unfinished' cathode materials containing nickel, cobalt and aluminium or manganese but not lithium—to some of the largest cathode materials makers both in China and South Korea.[28] Questions remain on how much of it comes from recycling activity. GEM's more recent deals gave the company access to freshly mined nickel from Indonesia[29] and cobalt from Congo,[30] as GEM outgrows its urban mining roots.

This ability to follow the money is somehow characteristic for Chinese organizations which tend to be started, expanded and led by the same entrepreneurial individuals. GEM is still led by a former university professor who specialized in recycling and who had founded the company in 2001.[31] Even if GEM markets itself with the slogan 'Recycling for future',[32] it did not hesitate to grow the battery materials business based on other sources than urban mining stocks. By some estimates, GEM is one of the top three biggest suppliers of refined cobalt to China's battery companies.[33]

This example of a recycling company becoming one of the battery materials market leaders is rare. Usually, it works the other way around. Huayou Cobalt, as the name suggests, is a cobalt producer. It owns a mine in Congo and refining facilities in China, with capacities that make the company the largest cobalt refiner in the world.[34] Huayou Cobalt also happens to operate one of the largest battery recycling facilities in China, with the ability to process tens of thousands of tons of battery waste per year. Brunp Recycling, acquired by the Chinese battery maker CATL, also has comparable battery waste capacities.[35] And there are other large recyclers. The point of this exercise, however, is to point out that large lithium-ion battery recycling capacities are already there. The lithium-ion battery recycling business is not something that may or may not happen in the future. It is happening as we speak, and the sheer scale of battery recycling capabilities in China is astounding.

If there is an issue for Chinese recyclers, it is the challenge of getting their hands on enough discarded batteries. But the strict regulatory regime, which obliges EV makers to take care of their discarded batteries, should soon turn battery collecting into an easy and streamlined process for recyclers. They will just have to liaise with EV makers to get their feedstock. In most of the recycling markets, the lack of centralized sourcing channels is a major problem. Electronic waste that would be valuable to recycle, for instance, often makes it to the back of the drawer or the attic instead of to the recycler.

R.I.D.—rest in drawer—became a recycling industry moniker for the universal phenomenon of no longer used electronic devices never reaching a recycler. Unwillingness to throw your no longer used laptop in the bin has a strong psychological grounding. It is easy to have concerns about personal data ending up in the wrong hands, and it might also be hard to discard an object on which one spent a considerable sum of money five years ago, and which, even if no longer up to standard, is still operational.

It would be interesting to see China's recycling landscape ten years from now. Will there be bidding wars of recyclers, with battery waste going to the highest bidder? Or maybe, on the contrary, automakers will have to pay up to dispose of their troublesome waste? Perhaps the government will step in and subsidize the recycling industry, just to make sure that the harvest of valuable metals takes place and that precious battery waste stays within the country. None of these conflicting scenarios can be eliminated as unlikely.

The concern remains for companies investing in large battery waste processing facilities that batteries will serve us longer than we expect. In China, you can already find innovative businesses offering battery regeneration for EVs. The fact that lithium-ion battery capacity fades is an unavoidable part of battery degradation, progressing with each additional cycle. But as EVs' huge battery packs are just collections of modules composed of single cells, it becomes easy to regenerate them. If the battery pack capacity degrades to 70 per cent, which typically makes it no longer usable, it is possible to replace the single most degraded cells within modules to bring the average pack capacity closer to 100 per cent. As it is unlikely that all the cells within the pack will degrade uniformly, this simple treatment could keep it going for much longer.

Another threat that may keep recyclers starving comes from the idea of battery reuse or second life. It is sound from an environmental and sustainability perspective to adopt a so-called recycling ladder. The concept was invented by a Dutch politician in 1979 and ranks waste management options from most to least environmentally desirable. Since re-using the object in a different capacity typically has less environmental impact, it makes more sense to attempt this first. So instead of treating a spent EV battery right away with fire and acids so as to harvest the metals in a recycling process, the environmental impact is lower if it is used for energy storage in an application that demands less performance than a modern vehicle.

In theory, the possibilities are endless: aggregated used EV battery packs could provide storage options for renewable energy sources such as wind or solar plants, or back-up power storage in hospitals or data centres. Such solutions could bring down costs and help the planet. While in Europe and the US numerous start-ups are gaining market traction with this idea, China already practises lithium-ion reuse on a grand scale, and with the help of eager state-owned telecommunication companies. Towards the end of 2020, China was close to fulfilling its target of rolling out half a million of its 5G base stations. Some of them will use second-life lithium-ion batteries as a back-up power storage. MIIT announced that this level of 5G station density puts Chinese 5G users at 60 million already.[36]

Reuse for the massive 5G network build-up already hoovers up a lot of discarded batteries from the market. As power back-up batteries they are unlikely to experience powerful discharging during acceleration or frequent cycling—which represented the typical stresses of their first lives as EV powertrains. The retired batteries will thus serve longer, and the metals inside them will have to wait to be recycled. An engineer testing the battery degradation levels told the anecdote that some of the batteries he deals with happened to 'heal' in their second life: their degradation level backtracked.

There are start-up technology companies who try to create a marketplace for used EV battery packs. Second-hand battery packs can fetch a respectable amount of money on the free market, with valuations between $5,000 and $15,000. There are platforms for used batteries on a model similar to eBay. The difference between regular auction sites and specialized EV battery auction sites is that the latter provide real time monitoring of battery parameters. The battery is still plugged in within the vehicle, which is being driven by the seller. You can see the battery's degradation level, mileage and cycle life right on the online platform. Where does this data come from? The seller does not need to install any kind of sophisticated chip that would collect and transmit this information. Such data is typically being sent to the automaker anyway, so the seller only authorizes the automaker to share it with the online platform. Reportedly premium EVs can transfer even 5GB of data per hour through their 5G-ready modems.

Information is power. The automakers need this data for R&D purposes to further improve their battery powertrains' performance. The software responsible for the data collection and real-time battery parameters control is called Battery Management System. It accounts for a battery brain of sorts. Some believe that the hardware side of the lithium-ion battery, at least in regard to existing cathode chemistries, is close to reaching its performance limits. You can, however, still squeeze in substantial improvements in battery performance by optimizing Battery Management Systems. But in order to do that, more data will be useful, and this is one of the reasons why automakers choose to monitor their EV fleet. With this level of monitoring in place, in theory carmakers could collect the information on an EV user's geolocation and driving behaviour. This prompts discussions about privacy concerns in regard of all this information.

Not all automakers will be happy to share their precious data, not only with online platforms selling used batteries but also with governments. Abundant and closely held data, after all, may provide them with a competitive advantage in this fast-growing industry. Behind closed doors, insiders say that this might be the reason why a tracking system for battery recycling and reuse may not work in some of the countries where the auto industry lobby is strong enough to influence legislative process in its favour.

Thus, the shape that lithium-ion battery recycling will take in the near future will be a product of a variety of factors—legislation, availability of waste, feasibility of its transportation between countries and the structure of a marketplace for it. It is comforting to know, though, that the technology to put battery waste somewhere else than in landfill is there and that it now depends purely on our choices on how and to what extent we will deploy it to protect the environment, while maximizing the economic benefit.

BRIGHT GREEN FUTURE

The fact that Greta Thunberg takes a train or even a yacht to reach climate conferences speaks volumes about the part aeroplanes and marine vessels play in the emissions of greenhouse gases. Aeroplanes contribute 2 per cent of all human-induced CO_2 emissions,[1] while the estimate for marine vessels is somewhere in the range of 2 to 3 per cent.[2] Whether it is a lot or a little, I leave to the reader to judge. Certainly, every action counts in the effort to stop climate change.

The idea to take the train for long-distance travel has seemed to really take off, and is leading to a small renaissance in railway travel. Railways started considering reopening long overnight connections between European capitals. The Scandinavians even coined terms such as 'flygskam' (flight shame) and 'tagskryt' (train bragging).

But can batteries help us make plane travel more environmentally sustainable? So that hopping off to Tenerife for a long weekend will not raise the eyebrows of some of our more environmentally conscious friends?

There are close to 200 electric aircraft projects globally, at different stages of development.[3] They range from small start-ups with drawing-board level concepts looking for funding to joint projects of big players such as Airbus and Rolls-Royce undertaking flight test characterizations. All these projects have one thing in common, though. They need to work around the limits of the current battery technology. At present, aeroplanes are powered by jet fuels, typically based on kerosene. In household applications kerosene is known as paraffin, the substance candles are made of.

Jet fuel is pumped into a fuel tank, which, in passenger aircraft, is situated in the wings. There are many advantages to such a solution—it reduces the stress on the wings, especially during take-off, by providing rigidity to what otherwise would be hollow spaces. It leaves space on the aircraft for cargo and, in the unlikely event of a crash landing, keeps it safely away from the passengers.

One good thing about jet fuel is that it reduces its weight during flight by simply burning away, thus extending the planes' ranges. But the best thing is the high level of energy density that it provides—meaning the amount of energy per unit of volume or mass. For jet fuels this is at around 12,000 Wh/kg. The maximum a lithium-ion battery can do in today's commercial applications is in the range of 250–300 Wh/kg—forty times less.[4] A Swiss start-up announced recently that they were developing a 1,000 Wh/kg battery, but the project is a long way from commercialization and definitely constitutes a cutting edge. If their technology was to make it to the market, it could extend the EV driving range to up to 1,000 km (620 miles) on a single charge. This is the kind of breakthrough we are talking about here. And yet it is still very far away from jet fuel's enviable 12,000 Wh/kg energy density.

So, at the current state of technology, lithium-ion batteries with the amount of energy equivalent to a full jet fuel tank would be so big and heavy that they would make flying impossible. Nevertheless, aviation companies take it one step at a time.

Technology almost always advances gradually. In the 1980s, it was difficult to fathom the computing power that processors were holding in 2020. It is unlikely that batteries will share the spectacular rate of development of semiconductors expressed in Moore's Law, which states that the number of transistors on a microchip will double about every two years, while their cost will halve. There is no Moore's Law for batteries. In 2010, cutting-edge lithium-ion batteries had a density in the range of 140–170 Wh/kg,[5] while in 2000, it was in the range of 120–150 Wh/kg.[6] In batteries, the pace of progress is less spectacular than it is in chips. But in the past five years, the pace of research on batteries has accelerated, and there has never been a time in history where it has attracted the amounts of brainpower and funding that it does today. Over the past three years alone, 53,000 academic articles on lithium-ion batteries have been published, not to mention papers on other battery technologies. As Henning Schoenenberger from Springer Nature, the renowned academic publisher, put it in the preface of the first ever AI-generated book: 'The future of mankind depends on progress in research on lithium-ion batteries'.[7]

As of today, there are three approaches to the problem of making flights electric. Companies may repurpose existing plane models by replacing fossil-fuel propulsion with an electric engine. Even though such a repurposed plane made history as the first purely electric flight, this method still has many limits. Only small planes the size of a Cessna can be repurposed. For larger aeroplanes, this simple approach will not work.

The second approach is to build hybrid planes, similar to hybrid cars in the sense of having both electric and fuel-powered engines on board. This is a strategy that large players in aviation such as Airbus are pursuing, replacing one of the four jet engines with an electric engine.[8] Hybrid solutions were the first to take hold on the roads and continue to be popular: they contribute to limiting emissions and to advancing battery technology. It seems that the energy revolution in the sky will follow a similar trajectory.

The third approach is to build all electric planes from the ground up, the way Tesla built its cars. Here, much as was the case with cars, start-ups seem to have pole position. The Israeli company Eviation is such a player. Co-founded and led by an ex-Israeli Defence Forces major with fifteen years of experience in the aviation and aerospace industry,[9] it has already built its first electric plane and is awaiting its test flight. 'Alice', as the plane model has been called, is a sleek machine with a luxurious interior, built to carry nine passengers on a flight with a range of 1,000 km (620 miles). Its design makes its being battery powered not the only selling point. What hits you first when you look at Alice is that it has propellers mounted in reverse on the wings, in a so-called pusher configuration. One pusher propeller is also mounted on the very end of the rear, reminiscent of a submarine. The unusual design is completed with a v-shaped tail wing, more often seen in military drones than in commercial aeroplanes. The energy comes from a 820 kWh, 3.7-ton NMC lithium-ion battery.[10] As a point of reference, Tesla's model 3 has a 75 kWh battery in its highest range version.[11] The battery is supplied by a small but long established South Korean specialized lithium-ion battery maker, which you most likely have never heard of—Kokam. As Eviation's CEO admitted, the number of cells they need at this stage would not justify starting a design process for a large battery maker such as LG Chem or Samsung SDI.[12] Kokam on the other hand had already made the batteries for the solar powered plane that pulled off the first battery-powered around-the-world flight. Alice's battery pack is so heavy at 3.7 tons that it accounts for 60 per cent of the plane's total weight. If it was put somewhere on board as one unit, it would likely make flying impossible. So the 9,400 NMC cells that make up the battery pack had to be spread out across the whole plane, including in spaces such as ceilings, floors and wings. Despite the significant size of the battery, Alice will take only three hours to charge.[13] A US regional airline operating on shorter routes in the Midwest and the Caribbean was Eviation's first customer. Other US airlines followed, and by the end of 2019, the order backlog for Alice reportedly amounted to 150 units.[14]

It is not only the care for the environment that makes flying electric a feasible choice. Savings are also considerable—running costs are estimated at $200 per hour as opposed to $1,000 for a conventionally powered turboprop used at similar ranges.[15] Electric planes are also eerily quiet—something that makes a difference for those living under flight paths and near airports. Alice's first flight had been planned for 2020, with introduction to the market in 2021. However, the aeroplane caught fire during the ground tests in Arizona's Prescott airport in January 2020. According to the company, the fire is believed to have been caused by a fault with the ground-based battery system.[16] The history of aviation is a history of constant improvement through trial and error. The Wright brothers would have never made it into the history books had it not been for their tenacity in the face of many unsuccessful attempts over the years. We cannot expect, despite the gigantic leap in technological advancement, that the road to flying electric will be free of potholes.

In 2017, it was announced that Airbus, Rolls-Royce and Siemens would partner on E-Fan X, a hybrid electric aircraft built on the base of BAe 146, a successful short-haul aircraft model produced in 387 units between 1983 and 2002.[17] BAe 146, in one of its versions, was able to carry more than 100 passengers on a range of 3,340km. Due to the number of units sold it has been called 'the most successful British civil jet airliner programme'.[18] Under the E-Fan X development, one of its four jet engines was replaced by Siemens' 2 MW electric motor, powered by a 2-ton, 700 kWh lithium-ion battery. The role of the battery in this hybrid system was to boost the power during take-off and climb, and to allow for electric-only descent that would lower the noise and emissions in airport areas during landing.[19] After extensive testing, E-Fan X was due for its first flight, but the programme was cancelled in 2020, amid the coronavirus pandemic. The companies involved asserted that the programme had been a stepping stone, which had allowed them to learn a lot about electric aviation, and that the effort had not been wasted.[20] But is there an electric plane, fully battery powered or hybrid, which already made it to the sky?

In December 2019, the world's first all-electric plane took off into the air from a lake in Vancouver. The flight lasted four minutes, during which the seaplane travelled 16km over the surface of the water. Despite the short length and the caution exercised by flying over the water, the event had a historical dimension. The de Havilland Beaver seaplane model was retrofitted with a one-ton lithium-ion battery and a 750 horsepower electric motor. The Canadian seaplane operator Harbour Air that operated the maiden flight hopes to electrify its entire fleet by 2022.[21]

Not long after, in May 2020, the feat was repeated, this time with Cessna Caravan 208B, which normally takes up to nine passengers. For the whole thirty minutes, the plane flew around Grant County International Airport in Washington. The electric motor was made by the same company that had equipped the Beaver—magniX.[22]

It cost only $6 to fly the electric Cessna for that long, while for the conventionally powered flight, it would have been around $300–$400. Those flights are still a very long way from flying electric from London to Japan. But the future of battery-powered planes on shorter, regional routes, especially in areas with an unspoilt natural environment, is closer than most would think. The low exploitation costs may also make learning to fly more affordable and thus more available to everyone.

The saltwater crocodile is the largest living reptile, with male representatives of the species weighing up to one ton and growing to 6m in length. They are also known as estuarine crocodiles, as in the past they bred in estuaries of South East Asian rivers, such as the Chinese Pearl River Delta.

Sadly, you will not find them anymore in the Pearl River Delta, as the area turned from a mostly agricultural region to the world's largest continuous urban centre. The region where the Pearl River flows into the South China Sea probably experienced the fastest urbanization in human history. In fact, it is continuing still, as China has announced plans to turn Guangzhou (population around 12 million), Shenzhen (around 9 million), Dongguan (6.5 million), Zhaoqing (3.9 million), Foshan (5.4 million), Huizhou (3.9 million), Jiangmen (3.8 million), Zhongshan (2.4 million) and Zhuhai (1.5 million) into one single megacity. These multi-million cities are connected by a network of waterways on which the freight volume reached 1 billion tons in 2020. This is the environment in which China decided to deploy the world's first large fully electric cargo vessel. It was built in Guangzhou's shipyard in 2017 and is 70.5m long—if placed vertically, it would reach as high as a twenty-story building. It can take up to 2,000 metric tons of cargo, which is equivalent to 80–100 full container loads. It is powered by a 2,400 kWh lithium-ion battery, and after a recharge of only two hours can travel a distance of over 60km, while fully loaded. The irony is that China's first zero-emission electric vessel is being employed to carry polluting coal to the power plants in the Pearl River Delta.[23]

Norway boasts the highest number of EVs per capita.[24] It is no wonder that it does not want to lag behind regarding electrified shipping solutions. Already in 2015, Storting, or 'The Great Assembly' as the Norwegian parliament is called, passed legislation requiring all national ferries to become electric with financial support from the state. From 2026 onwards, you will not be able to see fjords from fuelled vessels, as Norway's fjords will become the world's first zero emission zone at sea. Yara, a Norwegian chemical company, mostly known for its fertilizers, went a step further, aiming to build a cargo vessel that will be not only fully electric but also autonomous. Work on the project started in 2017, and Kongsberg, a technology company working on complex engineering solutions deployed from deep seas to outer space, was hired to build the vessel. Its voyages along the Norwegian coast, with a loading capacity of 120 containers, are supposed to save emissions from 40,000 truck journeys a year.[25] Yara Birkeland, as she is named, was expected to make her first unmanned sailing operation in 2022, but unfortunately the project has been paused during the pandemic.[26] Since the E-Fan X hybrid aeroplane encountered a similar fate, this chapter illustrates how many of Europe's most technologically cutting-edge projects lost priority during the recent difficult and uncertain times.

It may seem that at least in the case of the vessels the weight and size of the battery is a less sensitive issue than it is in planes. This, in theory, could help large electric cargo vessels become reality more easily than commercial battery-powered passenger aviation. To an extent, this statement is true, but we still cannot treat the challenges that batteries need to overcome to replace fuelled engines lightly. Today's cargo vessels carry tens and hundreds of thousands of tons of cargo between continents. The journey from the port of Shanghai to the port of Rotterdam is nearly 20,000km long, and cargo ships can make this journey without refuelling in between. Will this distance ever be possible on a single or a double battery charge? On the other hand, a large size containership burns 225 tons of bunker fuel a day on average during such journey, and it can easily keep 16,000 tons of fuel in a tank.[27] Engines on such vessels can be as high as four-story buildings and weigh easily over 2,000 tons. So, if we were to replace all this space and weight taken up by fuel and the fuelled engine with an evenly-distributed huge battery pack and a smaller electric engine, then perhaps we would have enough power to move the cargo vessels across the distances that would justify their use? I am assuming a smaller electric engine, as so far in cars and aeroplanes electric engines have turned out to be smaller as a result of fewer moving parts than in their conventional counterparts.[28] This, of course, is purely a thought experiment, and we need to leave it to engineers to assess the viability of such an idea.

Surely in case of such a monstrous battery pack, with high energy density, safety would be one of the major concerns. Catching fire from a thermal runaway in a battery is not an option, neither when airborne nor on the high seas.

The Canadian Sterling PlanB Energy Solution is developing safety solutions that will, it is hoped, make having a Plan B, in case of the ship catching fire, redundant. The company offers liquid cooling systems, where each cell is enclosed in its own cooling channel, with water moving around it, dissipating heat.[29] Even if water and electronics do not mix well, water has been successfully used to cool supercomputers, so perhaps now its time for cooling batteries has arrived.

The military often pushes the frontiers of innovation. It operates on large budgets, and the pressure to achieve a technological and thus strategic advantage over the potential enemy is enormous. Not all of the army's innovations are made public, but Japan decided to reveal its hybrid electric submarine, powered by lithium-ion batteries, to the world. 'Big Whale', or Taigei in Japanese, is the largest submarine Japan has built since the Second World War. It can take a crew of up to seventy on its underwater journey. Switching into battery-powered propulsion makes the submarine exceptionally quiet and difficult to detect for opponents. The range that it can travel on a lithium-ion battery, however, is confidential.

But in order to really turbo-charge the development of electric planes and vessels and alleviate anxiety about range once and for all for EV drivers, we need to move beyond lithium-ion chemistry. Moving away from lithium-ion does not necessarily mean moving away from lithium, however.

But how do you go about building a better battery? First, we need to remember that a battery is a closed system. In every closed system, if you change one component, you need to carefully consider the impact this has on the other components. This is true even if the change means an improvement. It seems common sense, but this rule is often forgotten with surprising frequency when breakthroughs in components development are enthusiastically being announced. You have a better cathode? Great, but the question remains how the better cathode will work with an existing electrolyte.

Seen from a higher level, the battery is remarkably simple. It has only three key elements: cathode, anode and electrolyte. If you want to improve aspects of its performance, such as its energy density, power, speed of charging or safety, first and foremost you will focus on making elements in this trio better. Of course, the field is so competitive that companies work on improving even the other, more peripheral elements of the battery to achieve marginal gains. But the biggest breakthroughs will entail only these three elements.

So we know that we have three elements to work on, and that, if we change one element, we will probably have to tweak the other two, to accommodate the changes and keep the balance within the system. But where to start improving? Do we at least know in which direction we want to go with our research and development?

In order to answer that, we need to go back to basic, high-school level chemistry. We may still remember that metals are eager to lose electrons and non-metals are willing to accept them. This process is a natural phenomenon, one that just needed to be discovered. So, if one element is losing electrons and the other is accepting them, this means that electrons are moving, and, as we remember, the movement of electrons means electricity. So, we are definitely on to something here. Further, the process of losing and gaining electrons is called ionization. The atom is in a neutral electric state if it has an equal number of electrons and protons. But if you put a metal and a non-metal together in a battery, ionization occurs: the metal loses electrons and the non-metal gains them. Atoms that are thrown out of balance and have more or fewer electrons than protons are called ions. So the term lithium-ion battery should make more sense now. You can have negative ions and positive ions. Negative ions are those atoms which have more electrons than in their electrically neutral state, and positive ions are atoms which have fewer electrons than in their electrically neutral state. Electrons always have a negative charge, so it is easy to remember that if you add a negative electron to an atom which is in an electrically neutral state, you turn the atom into a negative ion.

With that out of the way, to build a great battery we need to consider which elements from the periodic table that are happy to shed electrons, and elements that are willing to gain electrons, we need to put together to achieve the best performance. This is purely a theoretical exercise, but an immensely important one, because it shows what level of battery performance can be achieved just when factoring in the basic laws of chemistry and physics.

It is also a really practical one, because it demonstrates why we use lithium in the battery now, why we are interested in putting more lithium into the battery of the future, and why companies and researchers are focused on developing sulphur or sodium batteries for now, instead of other, metals-based batteries. After all, there are quite a few metals in the periodic table to choose from.

But to make laptops work on a battery for days, to make EVs go further on a single charge, to be able to one day fly electric from London to Barcelona on a passenger flight, we need energy dense batteries. This is the holy grail of battery research. Energy density is expressed as the ratio of stored energy to battery mass or battery volume. And in order to have a high energy density and low weight, we are limited to choose metals and non-metals from the top rows of the periodic table, as this is where the elements with lighter atomic mass sit. We also need ions that will shuttle back and forth, storing charge, and here our choice for elements is even more limited.[30] It is important to remember that when we speak about energy density, some elements may have a higher mass energy density and a lower volume energy density and vice versa in relation to other elements. This fine distinction matters a lot, because in the case of cars, where space is a more constraining factor than weight, it is volumetric energy density that matters.[31] Cars are of a certain standardized size, and you want the battery to take up as little of the volume of the car's interior as possible to have space for passengers and a trunk. On the other hand, in aeroplanes, energy density per unit of mass matters more than per unit of volume, as you want to take off into the air with the lightest battery possible. This distinction on the level of the elements in the periodic table also shows that it is difficult, if not impossible, to work on an ideal universal battery. There is no one size fits all here, and we need to look at the usefulness of different battery types toward specific applications. This also proves why it makes business sense to develop battery chemistries that may not be the best for EVs or aeroplanes, as they might just be fantastic for powering satellites in extremely cold outer space.

So as for weight lithium does very well with a very low atomic weight of 7. For the sake of comparison, lead used in lead acid batteries to start the ignition in fuel-powered cars has an atomic weight of 207, while the element sodium, which is often mentioned in the context of batteries of the future, has an atomic weight of 23. This is still very low, which makes it a good candidate for use in batteries, but nonetheless slightly higher than lithium. Before we go further, why do we even consider sodium for the battery of the future if it is heavier than lithium? Marginal loss on energy density perhaps does not make sodium an ideal candidate for the battery to power an aeroplane. But other factors matter here as well: sodium is much cheaper than lithium and more ubiquitous (it is an ingredient of common table salt after all), so it ticks the boxes as far as price and supply security are concerned.

You also need elements that would generate a high voltage. Electric power equals voltage multiplied by current. Accelerating an electric car from 0 to 100km/h necessitates a lot of electric power released in a short period of time. At present, electric cars already accelerate faster than fuelled vehicles. To check elements that, when put together, can generate a high voltage, one needs to switch from looking at the periodic table to the electrochemical potentials table. There, lithium again excels, offering a maximum of 4.5 V. If we look again at the old lead battery chemistry, lead electrochemical potential is below 2.1 V. Sodium also does very well, at a maximum of 4.2 V.

Using the Faraday constant and the molecular weight of electrochemically active materials, we can also calculate the theoretical capacities of the cell. We will not do that here, as the number of book sales is inversely proportional to the number of equations in it. It is pretty straightforward, though.

The capacity of the cell is the number of electric charges used through time that a battery can hold per unit of mass. It might be a tricky concept to grasp at first. This is because this unit of battery performance measure takes into account time, electric current and mass and encapsulates them into one number. To understand it we need to break it down. The relation between time and electric current is such that the same amount of electric current stored in a battery will last longer or shorter depending on how much current we will draw from it per hour or other unit of time. Some electric devices will obviously draw more and some will draw less current—just think of a fridge compared with a smart watch. Now you need to take this measure of electric current per time and contrast it with the volume of mass where you can store it. When you do this, you get theoretical values such as 3860 mAh/g (milliamp-hours per gram) for lithium, 4200 mAh/g for silicon, 1670 mAh/g for sulphur, or 370 mAh/g for graphite. These theoretical values for capacities are very, very high, even if we talk about graphite, which pales in comparison to lithium or sulphur, and we do not use them in full even in cutting-edge batteries. We would love to get close to theoretical capacity values—only there are many limiting factors that prohibit us from doing so. Building a better battery, a battery of the future, is to a large extent a struggle to overcome these limiting factors. It is a good thing that even a basic understanding of the electrochemistry guides us into directions in battery development which are worth exploring. For instance, if we know that lithium has high electrochemical potential, is light and has a very high theoretical capacity, we may want to use it to create a purely lithium-made anode, replacing the graphite anode that we have now. The same goes for silicon, or for sulphur or for sodium. Fundamental chemistry and physics let us understand that these elements, due to their properties, may find a place in the battery of the future to make it more energy dense.

There is an inherent beauty in the realization that a bunch of not too complex equations provides us with a detailed roadmap for the future. We know that with lithium sulphur batteries, using lithium metal as an anode and sulphur as a cathode, we will theoretically be able to get to 2,000 Wh/kg energy density, and with lithium oxygen batteries to 3,000 Wh/kg density. Theoretically means that in real life we will never have lithium sulphur batteries with 2,000 Wh/kg energy density, but if we find a way to develop and commercialize a lithium sulphur battery, we will get somewhere near this threshold. How close, we do not know. For that matter, we do not even know if the lithium sulphur battery, which is now under development, will ever fulfil the requirements in terms of cyclability (number of charges and discharges possible before degradation) that will qualify it for mass market use. We do not know that, but a number of companies invest resources in the belief that one day it will.

We throw these theoretical energy density numbers around, looking at what is imaginable in battery space. It is important to realize how large they are. The Tesla Model 3 is around 250–260 Wh/kg at cell level, and Elon Musk in one of his tweets in 2020 hinted that 400 Wh/kg is likely to enter the market within three to four years.[32] The Battery500 consortium, launched in 2017 and made of top universities and laboratories, as its name suggests targets the creation of the 500 Wh/kg battery.[33]

Since we are on the topic, there are two important things to understand about energy density. The first is how it is measured: what, really, is a 'Wh' or watt-hour? The easiest way to think about it is with the car analogy. In physics, power multiplied by time equals energy. Watt is a measure of power, and hour represents the time within which this power is applied. Think about a car travelling a certain distance, say 100km. You may travel those 100km faster or more slowly depending on how much you step on the gas—or in other words, how much power you use. In the case of the battery, you can imagine distance as time and speed as power. If you have 1kg of cells with 100 Wh/kg energy density and plug them to a fridge with a power rating of 100 Watt, you will have enough energy in these cells to power the fridge for one hour. If you double the weight of the cells to 2kg, you will have enough energy to power the fridge for two hours. If you take a bigger or less energy efficient fridge with, say, a 200-Watt rating, your 1kg of cells will last you only half an hour.

The second important point is that in battery world, we always discuss energy density at single cell and battery pack level. A battery pack powering an EV can be made of thousands of single cells. Since energy density is a measure of energy per unit of mass or volume, on the single cell level, energy density will always be higher than on the battery pack level. This is because battery packs may include connectors, cables, sensors and cooling mechanisms, which substantially add weight and volume. Since components such as cables and sensors do not store any energy, they are treated as dead weight. They add to the mass but do not add to energy storage, and thus decrease energy density. To give an idea of the gap between cell level energy density and battery pack energy density, we already mentioned that on cell level, for the Tesla Model 3 it is somewhere around 260 Wh/kg, while at battery pack level, it is around 160 Wh/kg, so a whole 100 Wh/kg lower.[34]

While energy density decreases from cell to battery pack, battery cost as measured in $/kWh (kilowatt-hours) increases as we move from cell to battery pack. One kilowatt-hour equals 1,000 watt-hours—we discussed energy density in watt-hours so far, as this is how it is typically expressed. Battery cost, on the other hand, is discussed in dollars per thousands of watt-hours, or kilowatt-hours. This price increase between cell and battery pack level is also only logical. When you look at pack level, you count in the costs of the elements that constitute the battery pack, so that again includes cables, connectors, sensors etc., while at the cell level you just look at the cell, without the extras. The battery cost differs a lot from manufacturer to manufacturer and also depends to a large extent on the type of lithium-ion battery chemistry used, and the type of raw materials that make up this chemistry. Larger proportions of more expensive raw materials such as cobalt or nickel drive the cost up, while the use of cheaper material in cathode chemistry, such as iron phosphate, brings it down. If we talk about the battery's future, we also need to consider that for a long time, the goal was to bring EVs' battery cost at pack level below $100/kWh. For some lithium-ion cathode chemistries such as LFP this has probably already happened,[35] while for high nickel cathode chemistries, the best performers stay somewhere in the $150/kWh territory. The goal of $100/kWh is so important because some believe this would bring EVs to cost parity with fuelled vehicles.[36]

Having discussed how we might make a better battery from the perspective of fundamental science, we should now have a closer look at how the single components work and how to make them better. As we already said, the focus is on cathode, anode and electrolyte. Both cathode and anode store lithium, while the electrolyte paves the way for lithium to pass from cathode to anode and back. Before the battery is charged for the first time, in the battery production process, lithium is inserted into the cathode. The cathode has a very interesting structure. It is made of polymer binder (a glue keeping it all together), carbon black (enhancing electrical conductivity) and the most important ingredient—intercalation oxide.

If you were to take an electron microscope and zoom in on the intercalation oxide, you would find that it is a crystal structure, with some empty spaces, or slots. These empty spaces make it possible for lithium to enter and leave. It enters during the discharging of the battery and leaves during charging. The structure of the crystals depends on the cathode material that we use. The LFP cathode has an olivine structure—with a huge stretch of the imagination, it resembles an olive tree with dense convoluted branches where lithium atoms are hiding. NMC and NCA have layered structures, like bookshelves on the wall hanging one over the other, but instead of books they hold lithium atoms. Even if

we operate at nano level here, the basic laws of physics, or—if we
dare to put it this way—common sense, still hold. For a battery
to quickly charge and to quickly discharge when we push the gas
pedal in an electric vehicle, lithium needs to be able to leave and
enter these structures quickly. So, to make batteries that allow
for an even faster charge, we need to make sure that the route for
lithium to travel into these structures is as short and as easy as
possible, with no obstacles on the way. We also need to ensure
that these crystal structures do not degrade when we insert and
take out lithium from them, thousands of times, during each
charge and discharge of the battery. When you insert lithium
atoms into the structure, it is reasonable to expect that such a
structure will expand. The same applies to the anode—it also
needs to accommodate lithium. It, too, degrades after many
insertions and extractions of lithium from its structures, expanding during insertion and contracting after lithium is removed. Of
course, it all happens very fast, and on an extremely micro scale.
We see the battery as a stable, immovable object, but inside it,
the chemical reactions continue, and electrodes expand and contract, while the battery breathes, almost like a living organism
taking in and releasing oxygen.

This is very important, because it puts a limitation on batteries
of the future. For a battery to be more energy dense, we would
have to be able to store more lithium in the cathode and the
anode. That seems easy—we turn to our roadmap, which we
have built on our understanding of basic electrochemistry. For
instance, we could build a sulphur cathode. It would bring our
costs down, as sulphur is a cheap and abundant material, and our
energy density up, as a sulphur cathode would be able to store
much more lithium than the most optimized NCM cathode.
Why, then, don't we do that?

Well, we could put such a battery up in an experimental setting, but when we charge and discharge it, it will quickly become

evident that the sulphur cathode eats up so much of the electrolyte, the liquid substance between cathode and anode, during the process, that the electrolyte quickly dries up and the battery stops working after only a few cycles. Thus, in theory, yes, we could make it work, by providing a huge storage for the electrolyte with it. But such storage would make the battery huge, so our goal of high energy density per weight and volume would be lost. You may think that an electrolyte problem cannot be impossible to fix—surely geniuses toiling away in the lab will one day overcome it. And this may be true, one day, but this problem has been known already for decades, and so far, nobody has solved it.

Further, if we can store much more lithium in the cathode, and lithium shuttles back and forth between cathode and anode, we need to make sure that the anode will be able to store this additional lithium.

For anodes, we have two great candidates to replace the graphite. Again, both come from our roadmap based on basic chemistry and physics. They are silicon and lithium. In real life though, both present a set of challenges. We said that a battery breathes, and the graphite anode expands by some 5 per cent when taking in lithium. Silicon will expand by 300–400 per cent.[37] To have high energy density, a battery needs to be tightly packed. If you expand one of the elements in a tightly packed closed system by three to four times, it is obvious that it will break. Again, it looks like a tractable problem. Scientists have been working for many years to solve it, though, and yet have not been able to commercialize a silicon anode on a large scale. One of the proposed solutions is to use silicon as an additive to existing graphite anode technology. Since one carbon atom can host only one of lithium, while one silicon atom can host four to five lithium atoms, it is enough to add up to 10 per cent of silicon to a graphite anode to increase its capacity dramatically, while keeping the expansion level under control.[38]

Using lithium metal as an anode is another interesting possibility. In fact, the first rechargeable lithium battery, invented by 2019's Nobel laureate Stanley Whittingham in the 70s, used lithium metal as an anode. The battery was not commercialized by Exxon though, which was assigned the patent, because it did not function well enough and safely enough.[39]

It is hugely ironic that the first lithium battery patent was assigned to an oil company, and that the technology that was used as the cornerstone of the battery revolution and then discarded is also the technology that holds such big promise for the future.

Later on, researchers figured out that graphite is a more practical choice for hosting carbon, even if a graphite anode has a several times lower capacity to store lithium than a lithium metal anode.

So, what problem does a lithium anode cause? It is time to introduce dendrites on the stage.

Under an electron microscope, dendrites look like thorny bushes springing up from the flat anode surface. If they grow high enough, they pierce the separator, the porous plastic membrane immersed in electrolyte that lets lithium ions pass but blocks electrons. Thus, they cause a short circuit, electrons start to flow not as intended, the battery heats up, oxygen starts to release and we have a fire or even an explosion.

The first idea to stop dendrites from growing was to introduce a solid electrolyte. It was a logical step to take, as lithium is a very soft metal. In fact, lithium in its metallic state is more like plasticine than like the metals that we meet in everyday use. However, lithium in its metallic state is not to be confused with lithium carbonate, hydroxide or spodumene concentrate, as *prima facie*, when one looks at their form and substance, they seem to have nothing to do with each other.

Lithium in its metallic state is also hyper reactive and very light. If you put a piece of lithium metal on water, it will first float on it (as it is lighter), and then it will explode.

In rechargeable lithium-ion batteries we currently use liquid electrolytes, which are composed of organic solvents, lithium salts and additives that enhance their desirable properties. Such a mixture is highly flammable, but scores relatively well if we look across the board at the characteristics desired in an ideal electrolyte. These are: strong ionic conductivity, no electric conductivity, no reactivity with cathode and anode materials, ability to withstand high voltage, low price and safety. Safety, it turns out, is actually one of the weakest points of today's electrolytes, which, as their Achilles heel, jeopardizes the level of safety of the whole battery.

Therefore, the motivation to develop a solid-state battery was two-fold—to contain the dendrite growth and to improve overall battery safety. For a solid electrolyte, ceramic compounds are mostly being proposed, as they provide high conductivity for lithium-ions.

Surprisingly, the idea fails so far in respect of dendrite growth. It is counterintuitive to expect a soft metal such as lithium to destroy ceramic compounds that are as hard as glass, but it turns out that this is exactly what is happening. But then, we all know that 'constant dropping wears away a stone'.

The same happens with dendrites: despite their softness, they can generate great pressure which cracks the ceramic, solid-state electrolyte.

That said, when you look at press announcements describing advances in solid-state batteries, it is important to ask whether they talk about solid-state with a lithium anode or without. For some reason, market observers tend to equate the solid-state battery with the lithium anode battery. These two are, however, different things. Commercialization of solid-state rechargeable batteries would already be a big achievement—it would make batteries exceedingly safe. They are already safe in the sense that the accident rate is estimated at several per billion. This is where sophisticated battery management systems controlling batteries in real time, as well as robust quality control, took us. Nevertheless, the flammable liquid is still there, constituting a potential danger.

The solid-state electrolyte alone, though, will not increase battery energy density. Only solid-state batteries with enhanced anode and cathode would be able to.

We live in the wake of the battery's golden age. We somehow got there, starting from Volta's first battery—a pile of copper and zinc plates separated by cardboard soaked in brine. If we think about it, Volta's 1799 invention is, on a macro level, not so different from modern-day batteries. The invention fascinated Napoleon to the extent that he maintained relations with Volta and made him a count. The history of the battery is thus inseparably bonded with the history of an electric age. After all, it was Volta who refused to give credit to Galvani's proposition of 'animal electricity'.

In the past thirty years, as humanity discovered and focused on lithium-ion chemistry, the development of the battery has accelerated. The advent of portable computers and other digital electronic devices gave the first impetus to push the limit of what is possible in the battery world. This was only magnified by growing awareness of climate change and risks involved, and further advanced by companies that made electric cars usable and even cool. It is likely that dreams of flying electric will keep the momentum in battery technology development. Never in history has so much talent and money been invested in batteries as it is now. Nonetheless we should continue to dream electric and work hard to fulfil those dreams leading to the green bright future.

ACKNOWLEDGEMENTS

Even if you write a book on something as niche as lithium and batteries, you still stand on the shoulders of giants in their respective fields. This book draws on the research and publications of such renowned experts, journalists and scholars as: Víctor Cofré, Joe Lowry, Prof. Judd C. Kinzley, Prof. Yet-Ming Chiang, Prof. Ying Shirley Meng, and many, many others.

Mr Lowry popularizes and chronicles the lithium industry as nobody else does, and my own understanding of the industry, as well as the book, benefited from that. Victor Cofré's biography of Julio Ponce Lerou was instrumental for my understanding of the history of the Chilean lithium sector. Prof. Judd C. Kinzley's work on Xinjiang was highly novel and eye-opening, and gave me a better grasp of the natural resource industry development of the region in its historical context. Lectures by Prof. Yet-Ming Chiang and Prof. Ying Shirley Meng were outstandingly helpful in explaining the inner workings of the battery without a doctoral-level qualification in electro-chemistry or nano-engineering.

I would also like to thank my publisher, Michael Dwyer, for giving me this opportunity, for putting his trust in a first-time author, and for recognizing the need for a book on this subject. I am also grateful to my editor, Maren Meinhardt, for the patience required to edit my manuscript.

ACKNOWLEDGEMENTS

I would also like to thank Sophie Gillespie, who helped me to edit the first drafts of the chapters, when the book was in its initial stages. Last but not least, my gratitude goes to Jon Curzon, the most responsive agent an author can dream of, who continuously helps me to become a better non-fiction writer.

NOTES

INTRODUCTION

1. Biello, D., 2010. *Where Did the Carter White House's Solar Panels Go?* [online] Scientific American. Available at: <https://www.scientificamerican.com/article/carter-white-house-solar-panel-array/> [Accessed 10 March 2021].
2. Ridley, M., 2017. *Amara's Law.* [online] Rationaloptimist.com. Available at: <https://www.rationaloptimist.com/blog/amaras-law/> [Accessed 10 March 2021].
3. KAH, M., 2019. *Columbia | SIPA Center on Global Energy Policy | Electric Vehicle Penetration and Its Impact On Global Oil Demand: A Survey of 2019 Forecast Trends.* [online] Energypolicy.columbia.edu. Available at: <https://www.energypolicy.columbia.edu/research/report/electric-vehicle-penetration-and-its-impact-global-oil-demand-survey-2019-forecast-trends> [Accessed 25 March 2021].
4. Maugouber, D. and Doherty, D., 2019. *Three Shifts in Road Transport That Threaten to Disrupt Oil Demand | BloombergNEF.* [online] BloombergNEF. Available at: <https://about.bnef.com/blog/three-drivers-curbing-oil-demand-road-transport/> [Accessed 25 March 2021].
5. Vorrath, S., 2020. *"New normal" for electric vehicle range will be 500km, says Musk.* [online] The Driven. Available at: <https://thedriven.io/2020/07/23/new-normal-for-electric-vehicle-range-will-be-500km-says-musk/> [Accessed 25 March 2021].
6. Isdp.eu. 2018. *Made in China 2025 backgrounder.* [online] Available at:

<https://isdp.eu/content/uploads/2018/06/Made-in-China-Backgrounder.pdf> [Accessed 25 March 2021].

7. Team, T., 2016. *Volkswagen's Strategy 2025 Focuses On A Greener Future For The Company.* [online] Forbes. Available at: <https://www.forbes.com/sites/greatspeculations/2016/06/21/volkswagens-strategy-2025-focuses-on-a-greener-future-for-the-company/?sh=32b58 6735ffc> [Accessed 10 March 2021].

8. Jaskula, B., 2015. *Mineral Commodities Summaries 2015—Lithium.* [online] S3-us-west-2.amazonaws.com. Available at: <https://s3-us-west-2.amazonaws.com/prd-wret/assets/palladium/production/mineral-pubs/lithium/mcs-2015-lithi.pdf> [Accessed 25 March 2021].

9. Pillot, C., 2016. *The Rechargable Battery Market & Main Trends 2015–25.* [online] Nextmove.fr. Available at: <https://nextmove.fr/wp-content/uploads/2016/09/7.-C.-PILLOT-Avicienne-Energy-Evolution-du-march%C3%A9-mondial-des-batteries-pour-l%E2%80%99%C3%A9lectromobilit%C3%A9-Pl%C3%A9ni%C3%A8re-Moveo-VE_Mythes_et_R%C3%A9alit%C3%A9s.pdf> [Accessed 25 March 2021].

10. Okubo, M., 2019. *Creating a future energy world on the foundation of technology and innovation.* [online] The Japan Times. Available at: <https://www.japantimes.co.jp/country-report/2019/06/28/north-rhine-westphalia-report-2019/creating-future-energy-world-foundation-technology-innovation/> [Accessed 25 March 2021].

11. Sony.net. 2020. *Sony.* [online] Available at: <https://www.sony.net/SonyInfo/csr/SonyEnvironment/initiatives/pdf/2015_Chargeyour Emotion.pdf> [Accessed 25 March 2021].

12. Yuanyuan, L., 2020. *China installed more than 1000 EV charging stations per day in 2019—Renewable Energy World.* [online] Renewable Energy World. Available at: <https://www.renewableenergyworld.com/storage/china-installed-more-than-1000-ev-charging-stations-per-day-in-2019/#gref> [Accessed 10 March 2021].

13. Ambrose, H., 2020. *How Long Will My EV Battery Last? (and 3 Tips to Help it Last Longer).* [online] Union of Concerned Scientists. Available at: <https://blog.ucsusa.org/hanjiro-ambrose/how-long-will-my-ev-battery-last-and-3-tips-to-help-it-last-longer> [Accessed 25 March 2021].

14. Yu, J., Che, J., Omura, M. and B. Serro, K., 2011. Emerging Issues on Urban Mining in Automobile Recycling: Outlook on Resource Recycling in East Asia. *Integrated Waste Management—Volume II.*

1. CHINA: A TREND MAKER

1. Chinadaily.com.cn. 2012. *Xi highlights national goal of rejuvenation—China.* [online] Available at: <https://www.chinadaily.com.cn/china/2012-11/30/content_15972687.htm> [Accessed 9 March 2021].
2. Moshinsky, B., 2015. *Here's why China mentioned the word 'innovation' 71 times after a meeting to decide its 5-year plan.* [online] Business Insider. Available at: <https://www.businessinsider.com/chinese-government-said-innovation-71-times-after-a-meeting-to-decide-its-5-year-plan-2015–11?r=US&IR=T> [Accessed 9 March 2021].
3. Isdp.eu. 2018. *Made in China 2025 backgrounder.* [online] Available at: <https://isdp.eu/content/uploads/2018/06/Made-in-China-Backgrounder.pdf> [Accessed 25 March 2021].
4. Bremner, R., 2018. *The world's first plug-in hybrid car—and why it failed—Retro Motor.* [online] The world's first plug-in hybrid car—and why it failed. Available at: <https://www.retromotor.co.uk/great-motoring-disasters/2011-chevrolet-volt/> [Accessed 26 March 2021].
5. DiNucci, M., 2017. *The Complete Guide to Charging the Chevy Volt.* [online] ChargePoint. Available at: <https://www.chargepoint.com/blog/complete-guide-charging-chevy-volt/> [Accessed 26 March 2021].
6. AAA Foundation. 2019. *American Driving Survey 2014–2017.* [online] Available at: <https://aaafoundation.org/american-driving-survey-2014-2017/#:-:text=Key%20Findings,miles%2C%20in%202016%20and%202017.> [Accessed 10 March 2021].
7. Collins, G. and Erickson, A., 2011. *Electric Bikes are China's Real Electric Vehicle Story | China SignPost™.* [online] Chinasignpost.com. Available at: <https://www.chinasignpost.com/2011/11/07/electric-bikes-are-chinas-real-electric-vehicle-story/> [Accessed 9 March 2021].
8. Song Meanings and Facts. 2018. *Meaning of "Nine Million Bicycles" by Katie Melua—Song Meanings and Facts.* [online] Available at: <https://www.songmeaningsandfacts.com/meaning-of-nine-million-bicycles-by-katie-melua/> [Accessed 10 March 2021].

9. Yang, C., n.d. *Launching Strategy for Electric Vehicles: Lessons from China and Taiwan.* [online] Web.archive.org. Available at: <https://web.archive.org/web/20100331153729/http://www.duke.edu/-cy42/EV.pdf> [Accessed 26 March 2021].

10. Mann, J., 1997. *Beijing Jeep: A Case Study Of Western Business In China.* Westview Press, p. 149.

11. Harwit, E., 2001. The Impact of WTO Membership on the Automobile Industry in China. *The China Quarterly,* 167.

12. Mann, J., 1997. *Beijing Jeep: A Case Study Of Western Business In China.* Westview Press, pp. 151–152.

13. Bradsher, K., 2020. *How China Obtains American Trade Secrets (Published 2020).* [online] Nytimes.com. Available at: <https://www.nytimes.com/2020/01/15/business/china-technology-transfer.html?auth=login-facebook> [Accessed 9 March 2021].

14. Bloomberg.com. 2020. *The Solar-Powered Future Is Being Assembled in China.* [online] Available at: <https://www.bloomberg.com/features/2020-china-solar-giant-longi/?sref=TtblOutp> [Accessed 26 March 2021].

15. Hanada, Y., 2019. *China's solar panel makers top global field but challenges loom.* [online] Nikkei Asia. Available at: <https://asia.nikkei.com/Business/Business-trends/China-s-solar-panel-makers-top-global-field-but-challenges-loom> [Accessed 26 March 2021].

16. Tabeta, S., 2017. *Changan Auto sells 3m cars in record year.* [online] Nikkei Asia. Available at: <https://asia.nikkei.com/Spotlight/Auto-Industry-Upheaval2/Changan-Auto-sells-3m-cars-in-record-year> [Accessed 26 March 2021].

17. Dixon, T., 2017. *Changan Automobile aims to end sales of traditional fuel vehicles by 2025 and invest 100 billion Yuan (15 Billion USD) into electrification.* [online] EV Obsession. Available at: <https://evobsession.com/changan-automobile-aims-end-sales-traditional-fuel-vehicles-2025-invest-100-billion-yuan-15-billion-usd-electrification/> [Accessed 26 March 2021].

18. Qiao, Y., 2010. *Milestone merger reshapes Suzuki.* [online] Chinadaily.com.cn. Available at: <http://www.chinadaily.com.cn/bizchina/2010-03/29/content_9655056.htm> [Accessed 26 March 2021].

19. Reuters.com. 2009. *UPDATE 2-Changan Auto claims China's No. 3 spot with AVIC deal.* [online] Available at: <https://www.reuters.com/article/changan-avic-idUSSHA28311020091110> [Accessed 26 March 2021].

20. Duff, M., 2019. *Year of the underdog: Geely's rise from obscurity to the top | Autocar.* [online] Autocar. Available at: <https://www.autocar.co.uk/car-news/features/year-underdog-geelys-rise-obscurity-top> [Accessed 26 March 2021].

21. Chang, C., 2009. Developmental Strategies in a Global Economy: The Unexpected Emergence of China's Indigenous Auto Industry. *APSA 2009 Toronto Meeting Paper*,.

22. Fairclough, G., 2007. *In China, Chery AutomobileDrives an Industry Shift.* [online] Wall Street Journal. Available at: <https://www.wsj.com/articles/SB119671314593812115> [Accessed 26 March 2021].

23. Li, L., 2005. *GM Daewoo files suit against Chery.* [online] Chinadaily.com.cn. Available at: <http://www.chinadaily.com.cn/english/doc/200505/09/content_440334.htm> [Accessed 10 March 2021].

24. China National Administration of GNSS and Applications (CNAGA). 2019. *Chen Fangyun, a Man of Great Merit for China's Nuclear Bomb, Missile and Satellite Undertaking: Work Selflessly for the Prosperity of My Motherland.* [online] Available at: <http://en.beidouchina.org.cn/c/1356.html> [Accessed 30 March 2021].

25. Sigurdson, J. and Jiang, J., 2007. *Technological superpower China.* Cheltenham, UK: Edward Elgar, p. 43.

26. Ie.china-embassy.org. n.d. *HIGH TECH RESEARCH AND DEVELOPMENT (863) PROGRAMME.* [online] Available at: <http://ie.china-embassy.org/eng/ScienceTech/ScienceandTechnology DevelopmentProgrammes/t112844.htm> [Accessed 9 March 2021].

27. Gewirtz, J., 2019. 'The Futurists of Beijing: Alvin Toffler, Zhao Ziyang, and China's "New Technological Revolution," 1979–1991'. *The Journal of Asian Studies*, 78(1), pp. 115–140.

28. Innovationpolicyplatform.org. n.d. *SYSTEM INNOVATION: CASE STUDIES: CHINA—The Case of Electric Vehicles.* [online] Available at: <http://www.innovationpolicyplatform.org/www.innovationpolicyplatform.org/system/files/CHINA%20-%20The%20Case%20

of%20Electric%20Vehicles-%20IPP_0/index.pdf> [Accessed 30 March 2021].

29. Marsters, P., 2009. *Electric Cars: The Drive for a Sustainable Solution in China.* [online] Wilson Center. Available at: <https://www.wilsoncenter.org/publication/electric-cars-the-drive-for-sustainable-solution-china> [Accessed 30 March 2021].

30. Marquis, C., Zhang, H. and Zhou, L., 2013. *China's quest to adopt electric vehicles.* [online] Hbs.edu. Available at: <https://www.hbs.edu/ris/Publication%20Files/Electric%20Vehicles_89176bc1–1aee-4c6e-829f-bd426beaf5d3.pdf> [Accessed 9 March 2021].

31. Zhang, Z., 2020. *China's 46 New Cross-Border E-Commerce Zones: A Brief Primer.* [online] China Briefing News. Available at: <https://www.china-briefing.com/news/china-unveils-46-new-cross-border-e-commerce-zones-incentives-foreign-investors-faqs/> [Accessed 10 March 2021].

32. Marquis, C., Zhang, H. and Zhou, L., 2013. *China's quest to adopt electric vehicles.* [online] Hbs.edu. Available at: <https://www.hbs.edu/ris/Publication%20Files/Electric%20Vehicles_89176bc1–1aee-4c6e-829f-bd426beaf5d3.pdf> [Accessed 9 March 2021].

33. Yang, Y., 2020. *New energy vehicles comprise 60% of 2020 Beijing quota.* [online] Chinadaily.com.cn. Available at: <http://www.chinadaily.com.cn/a/202002/07/WS5e3d1667a310128217275d49.html#:-:text=This%20year's%20annual%20city%20quota,small%20passenger%20vehicle%20quota%20announced.> [Accessed 10 March 2021].

34. BloombergNEF. 2020. *Battery Pack Prices Cited Below $100/kWh for the First Time in 2020, While Market Average Sits at $137/kWh | BloombergNEF.* [online] Available at: <https://about.bnef.com/blog/battery-pack-prices-cited-below-100-kwh-for-the-first-time-in-2020-while-market-average-sits-at-137-kwh/> [Accessed 30 March 2021].

35. Marquis, C., Zhang, H. and Zhou, L., 2013. *China's quest to adopt electric vehicles.* [online] Hbs.edu. Available at: <https://www.hbs.edu/ris/Publication%20Files/Electric%20Vehicles_89176bc1–1aee-4c6e-829f-bd426beaf5d3.pdf> [Accessed 9 March 2021].

36. Innermongolia.chinadaily.com.cn. 2018. *Hohhot opens first EV charging station.* [online] Available at: <http://innermongolia.chinadaily.com.cn/2018–02/09/c_140989.htm> [Accessed 30 March 2021].

37. Pigato, M., Black, S., Dussaux, D., Mao, Z., McKenna, M., Rafaty, R. and Touboul, S., 2020. *Technology Transfer and Innovation for Low-Carbon Development*. Washington, DC, USA: World Bank Group Publications, pp. 101–103.

38. Wang, H. and Kimble, C., 2010. 'Betting on Chinese electric cars?; analysing BYD's capacity for innovation'. *International Journal of Automotive Technology and Management*, 10(1), p. 77.

39. Iris Quan, X., Loon, M. and Sanderson, J., 2018. *Innovation in the local context—A case study of BYD in China*. [online] Eprints.worc.ac.uk. Available at: <http://eprints.worc.ac.uk/4603/1/ABM15_paper_27.pdf> [Accessed 10 March 2021].

40. Einhorn, B., 2010. *The 50 Most Innovative Companies*. [online] Bloomberg.com. Available at: <https://www.bloomberg.com/news/articles/2010-04-15/the-50-most-innovative-companies?sref=TtblOutp> [Accessed 30 March 2021].

41. BloombergNEF. 2017. *China's Menswear Maker Swaps Stitching for Lithium Batteries | BloombergNEF*. [online] Available at: <https://about.bnef.com/blog/chinas-menswear-maker-swaps-stitching-for-lithium-batteries/> [Accessed 30 March 2021].

42. Ibid.

43. Iris Quan, X., Loon, M. and Sanderson, J., 2018. *Innovation in the local context—A case study of BYD in China*. [online] Eprints.worc.ac.uk. Available at: <http://eprints.worc.ac.uk/4603/1/ABM15_paper_27.pdf> [Accessed 10 March 2021].

44. Ibid. p. 7.

45. Ibid. p. 8–10.

46. Chinadaily.com.cn. 2010. *BYD plans to start European car sales next year*. [online] Available at: <http://www.chinadaily.com.cn/bizchina////2010–03/09/content_9559285.htm> [Accessed 31 March 2021].

47. Chinaautoweb.com. 2014. *"Plug-in EV Sales in China Rose 37.9% to 17,600 in 2013"*. [online] Available at: <http://chinaautoweb.com/2014/01/plug-in-ev-sales-in-china-rose-37-9-to-17600-in-2013/> [Accessed 31 March 2021].

48. Yeung, G., 2018. '"Made in China 2025": the development of a new energy vehicle industry in China'. *Area Development and Policy*, 4(1), pp. 39–59.

49. Berman, B., 2011. *BYD Is the First Ripple in a Potential Chinese Wave (Published 2011).* [online] Nytimes.com. Available at: <https://www.nytimes.com/2011/02/20/automobiles/autoreviews/byd-f3-dm-review.html> [Accessed 31 March 2021].
50. Kane, M., 2016. *BYD Qin Sales Top 50,000, Tang Exceed 30,000.* [online] InsideEVs. Available at: <https://insideevs.com/news/331035/byd-qin-sales-top-50000-tang-exceed-30000/> [Accessed 31 March 2021].
51. Chinacartimes.com. 2013. *BYD Launches Qin Plugin Hybrid— 189,800RMB to 209,800RMB.* [online] Available at: <https://web.archive.org/web/20131221190347/http://www.chinacartimes.com/2013/12/byd-launches-qin-plugin-hybrid-189800rmb-209800rmb/> [Accessed 31 March 2021].
52. Patil, P., 2008. *Developments in Lithium-Ion Battery Technology in The Peoples Republic of China.* [online] Publications.anl.gov. Available at: <https://publications.anl.gov/anlpubs/2008/02/60978.pdf> [Accessed 31 March 2021].
53. Dongmei, L., Binbin, Y., Duan, W. and Yanyan, F., 2010. *How Manufacturing's Mockingbird Sings.* [online] Caixinglobal.com. Available at: <https://www.caixinglobal.com/2010–02–10/how-manufacturings-mockingbird-sings-101018597.html> [Accessed 16 April 2021].
54. Energy Central. 2020. *World Battery Production.* [online] Available at: <https://energycentral.com/c/ec/world-battery-production> [Accessed 31 March 2021].
55. Manthey, N., 2018. *BYD to more than double battery production by 2020.* [online] electrive.com. Available at: <https://www.electrive.com/2018/06/27/byd-to-more-than-double-battery-production-by-2020/> [Accessed 31 March 2021].
56. Govardan, D., 2019. *Tianjin Battery Co to buy cells from Munoth Industries.* [online] The Times of India. Available at: <https://timesofindia.indiatimes.com/business/india-business/tianjin-battery-co-to-buy-cells-from-munoth-industries/articleshow/71318951.cms> [Accessed 31 March 2021].
57. Jacoby, M., 2019. *It's time to get serious about recycling lithium-ion batteries.* [online] Cen.acs.org. Available at: <https://cen.acs.org/materi-

als/energy-storage/time-serious-recycling-lithium/97/i28> [Accessed 19 March 2021].

58. Kinzley, J., 2018. *Natural resources and the new frontier.* University of Chicago Press, p. 150.

59. Ibid. p. 99.

60. Ibid.

61. Kinzley, J., 2012. *Staking Claims to China's Borderland: Oil, Ores and Statebuilding in Xinjiang Province, 1893–1964.* [online] Escholarship. org. Available at: <https://escholarship.org/content/qt3p7432md/qt3p7432md_noSplash_40b53d988fe164c1496cc2f5e35ad314.pdf> [Accessed 31 March 2021].

62. Kinzley, J., 2018. *Natural resources and the new frontier.* Chicago, USA: University of Chicago Press, p. 123.

63. Ibid. p. 139.

64. Ibid p. 143.

65. Xin, L. and Lingzhi, F., 2019. *Exhibition shows vital role Koktokay played in building country.* [online] Globaltimes.cn. Available at: <http://www.globaltimes.cn/content/1158175.shtml> [Accessed 9 March 2021].

66. Kinzley, J., 2018. *Natural resources and the new frontier.* Chicago, USA: University of Chicago Press, p. 157.

67. Ibid.

68. Ibid. p. 171.

69. Albright, D., Burkhard, S., Gorwitz, M. and Lach, A., 2017. *North Korea's Lithium 6 Production for Nuclear Weapons.* [online] Isis-online. org. Available at: <https://isis-online.org/uploads/isis-reports/documents/North_Korea_Lithium_6_17Mar2017_Final.pdf> [Accessed 31 March 2021].

2. STRUGGLE FOR GLOBAL DOMINANCE

1. Itdcw.com. 2018. 赣锋锂业董事长李良彬：打造锂产业链"A+H"样板_电池网. [online] Available at: <http://www.itdcw.com/news/focus/0Z3955032018.html> [Accessed 9 March 2021].

2. Ganfeng Lithium Co., Ltd. 2019. *Ganfeng Lithium Annual Report 2018.* [online] Available at: <https://www1.hkexnews.hk/listedco/listconews/sehk/2019/0424/ltn201904241477.pdf> [Accessed 1 April 2021].

3. Jxxy.com. 2014. *李良彬—赣锋锂业董事长*. [online] Available at: <http://www.jxxy.com/show.aspx?id=359&cid=45> [Accessed 1 April 2021].

4. Escn.com.cn. 2016. *赣锋锂业：锂材料巨头成长记 - 锂电池 - 中国储能网*. [online] Available at: <http://www.escn.com.cn/news/show-347689.html> [Accessed 9 March 2021].

5. Xianjichina.com. 2020. *赣锋锂业老总简介【赣锋锂业董事长李良彬】-贤集网* [online] Available at: <https://www.xianjichina.com/news/details_1180 53.html> [Accessed 1 April 2021].

6. Business.sohu.com. 2010. *赣锋锂业上市造富 诞生12名千万富翁*. [online] Available at: <https://business.sohu.com/20100803/n273949130.shtml> [Accessed 1 April 2021].

7. Ibid.

8. Lowry, J., 2018. *Ganfeng: Still Under the Radar?*. [online] Linkedin. com. Available at: <https://www.linkedin.com/pulse/ganfeng-still-under-radar-joe-lowry/> [Accessed 1 April 2021].

9. Lowry, J., 2018. *E25: Not Lost in Translation*. [podcast] Global Lithium Podcast. Available at: <http://lithiumpodcast.com/podcast/e25-not-lost-in-translation-%E6%B2%A1%E6%9C%89%EF%BC%89%E8%BF%B7%E5%A4%B1%E5%9C%A8%E7%BF%BB%E8%AF%91%E4%B8%AD/> [Accessed 9 March 2021].

10. Ibid.

11. Finance.sina.com.cn. 2016. *专访赣锋锂业董事长：尽快挺进上游产品线*. [online] Available at: <http://finance.sina.com.cn/roll/2016–09–21/doc-ifxvyqwa3654069.shtml> [Accessed 9 March 2021].

12. Lowry, J., 2018. *E25: Not Lost in Translation*. [podcast] Global Lithium Podcast. Available at: <http://lithiumpodcast.com/podcast/e25-not-lost-in-translation-%E6%B2%A1%E6%9C%89%EF%BC%89%E8%BF%B7%E5%A4%B1%E5%9C%A8%E7%BF%BB%E8%AF%91%E4%B8%AD/> [Accessed 9 March 2021].

13. Ibid.

14. Ibid.

15. Ibid.

16. Ibid.

17. Reuters.com, 2018. *China's Ganfeng buys SQM's stake in lithium project for $87.5 mln.* [online] Available at: <https://www.reuters.com/

article/sqm-ganfeng-lithium-idUSL1N1V500c> [Accessed 2 April 2021].

18. Lowry, J., 2018. *E25: Not Lost in Translation*. [podcast] Global Lithium Podcast. Available at: <http://lithiumpodcast.com/podcast/e25-not-lost-in-translation-%E6%B2%A1%E6%9C%89%EF%BC%89%E8%BF%B7%E5%A4%B1%E5%9C%A8%E7%BF%BB%E8%AF%91%E4%B8%AD/> [Accessed 9 March 2021].

19. Ober, J., 1995. *Lithium*. [online] S3-us-west-2.amazonaws.com. Available at: <https://s3-us-west-2.amazonaws.com/prd-wret/assets/palladium/production/mineral-pubs/lithium/450495.pdf> [Accessed 15 March 2021].

20. LinkedIn. n.d. *GanfengLithium*. [online] Available at: <https://www.linkedin.com/company/ganfenglithium/> [Accessed 2 April 2021].

21. Lowry, J., 2018. *E25: Not Lost in Translation*. [podcast] Global Lithium Podcast. Available at: <http://lithiumpodcast.com/podcast/e25-not-lost-in-translation-%E6%B2%A1%E6%9C%89%EF%BC%89%E8%BF%B7%E5%A4%B1%E5%9C%A8%E7%BF%BB%E8%AF%91%E4%B8%AD/> [Accessed 9 March 2021].

22. Business.sohu.com. 2010. 赣锋锂业上市造富 诞生12名千万富翁. [online] Available at: <https://business.sohu.com/20100803/n273949130.shtml> [Accessed 1 April 2021].

23. Ibid.

24. Quotes.money.163.com. 2010. 江西赣锋锂业股份有限公司首次公开发行股票招股说明书摘要. [online] Available at: <http://quotes.money.163.com/f10/ggmx_002460_584934.html> [Accessed 2 April 2021].

25. Ibid.

26. Ibid.

27. Jones, T., 2013. *Chinese millionaire fights pollution with thin air*. [online] Reuters.com. Available at: <https://www.reuters.com/article/us-china-pollution-cans/chinese-millionaire-fights-pollution-with-thin-air-idINBRE90T0LM20130130?edition-redirect=in> [Accessed 15 March 2021].

28. BBC News. 2017. *China does U-turn on coal ban to avert heating crisis*. [online] Available at: <https://www.bbc.com/news/world-asia-42266768> [Accessed 3 April 2021].

29. Court, M., Rutland, T. and Dhokia, K., 2019. *China And The Environment—Industry Versus Air.* [online] Spglobal.com. Available at: <https://www.spglobal.com/marketintelligence/en/news-insights/research/china-and-the-environment-industry-versus-air> [Accessed 15 March 2021].

30. Reuters.com, 2019. *Beijing set to exit list of world's top 200 most-polluted cities: data.* [online] Available at: <https://www.reuters.com/article/us-china-pollution-beijing-idUSKCN1VX05z> [Accessed 15 March 2021].

31. Itdcw.com. 2018. *赣锋锂业董事长李良彬：打造锂产业链"A+H"样板_电池网.* [online] Available at: <http://www.itdcw.com/news/focus/0Z3955032018.html> [Accessed 9 March 2021].

32. Ibid.

33. Ibid.

34. Ibid.

35. Ibid.

36. Sgs.com. 2010. *HARD ROCK LITHIUM PROCESSING.* [online] Available at: <https://www.sgs.com/-/media/Global/Documents/Flyers%20and%20Leaflets/SGS-MIN-WA109-Hard-Rock-Lithium-Processing-EN-11.pdf> [Accessed 6 April 2021].

37. Facada, M., 2018. *LSM 18: What we learned at the 10th Lithium Supply & Markets Conference.* [online] Fastmarkets. Available at: <https://www.fastmarkets.com/article/3817787/lsm-18-what-we-learned-at-the-10th-lithium-supply-markets-conference> [Accessed 6 April 2021].

38. Sgs.com. 2010. *HARD ROCK LITHIUM PROCESSING.* [online] Available at: <https://www.sgs.com/-/media/Global/Documents/Flyers%20and%20Leaflets/SGS-MIN-WA109-Hard-Rock-Lithium-Processing-EN-11.pdf> [Accessed 6 April 2021].

39. Sherwood, D., 2020. *Exclusive: Lithium giants Albemarle and SQM battle over access to Atacama water study.* [online] Reuters.com Available at: <https://www.reuters.com/article/us-chile-lithium-albemarle-exclusive/exclusive-lithium-giants-albemarle-and-sqm-battle-over-access-to-atacama-water-study-idUKKBN27X10g> [Accessed 6 April 2021].

40. Scheyder, E., 2018. *Inside Albemarle's quest to reinvent the lithium market.* [online] Reuters.com Available at: <https://www.reuters.com/arti-

cle/us-albemarle-lithium-focus-idUSKCN1LF0Bj> [Accessed 6 April 2021].

41. Industry Europe. 2021. *Aluminium groups call on G7 to cut back on subsidies.* [online] Available at: <https://industryeurope.com/aluminium-groups-call-on-g7-to-cut-back-on-subsidies/> [Accessed 6 April 2021].

42. Tradingeconomics.com. n.d. *Aluminum | 1989–2021 Data | 2022–2023 Forecast | Price | Quote | Chart | Historical.* [online] Available at: <https://tradingeconomics.com/commodity/aluminum> [Accessed 6 April 2021].

43. Deforche, F., Bradtke, T., Deniskin, R., Gilbert, M., Gruner, K., Olav, K., Harlacher, D. and Koch, A., 2013. *The Aluminium Industry CEO Agenda, 2013–2015.* [online] Image-src.bcg.com. Available at: <http://image-src.bcg.com/Images/The_Aluminum_Industry_CEO_Agenda_2013–2015_June_2013_tcm9-95118.pdf> [Accessed 6 April 2021].

44. Tradingeconomics.com. n.d. *Aluminum | 1989–2021 Data | 2022–2023 Forecast | Price | Quote | Chart | Historical.* [online] Available at: <https://tradingeconomics.com/commodity/aluminum> [Accessed 6 April 2021].

45. CM Group. n.d. *Primary Aluminium | CM Group.* [online] Available at: <https://www.cmgroup.net/industries/primary-aluminium/> [Accessed 6 April 2021].

46. Facada, M., 2019. *Global lithium supply developing at accelerating pace on growing demand.* [online] Metalbulletin.com. Available at: <https://www.metalbulletin.com/Article/3868440/Global-lithium-supply-developing-at-accelerating-pace-on-growing-demand.html> [Accessed 15 March 2021].

47. Ouerghi, D. and Shi, C., 2020. *GLOBAL LITHIUM WRAP: Bullish suppliers push China's carbonate price up 3.4%.* [online] Metalbulletin.com. Available at: <https://www.metalbulletin.com/Article/3967364/GLOBAL-LITHIUM-WRAP-Bullish-suppliers-push-Chinas-carbonate-price-up-34.html> [Accessed 6 April 2021].

48. Sanderson, H., 2017. *Electric car demand sparks lithium supply fears.* [online] Ft.com. Available at: <https://www.ft.com/content/90d65356–4a9d-11e7-919a-1e14ce4af89b> [Accessed 15 March 2021].

49. Www2.deloitte.com. 2015. *Smartphone batteries: better but no breakthrough.* [online] Available at: <https://www2.deloitte.com/content/

dam/Deloitte/global/Documents/Technology-Media-Telecommuni-cations/gx-tmt-pred15-smartphone-batteries.pdf> [Accessed 6 April 2021].

50. Lambert, F., 2016. *Breakdown of raw materials in Tesla's batteries and possible bottlenecks.* [online] Electrek. Available at: <https://electrek.co/2016/11/01/breakdown-raw-materials-tesla-batteries-possible-bot-tleneck/> [Accessed 15 March 2021].

51. 2019. *Ganfeng Lithium Annual Report 2018.* [online] Available at: <https://www1.hkexnews.hk/listedco/listconews/sehk/2019/0424/ltn201904241477.pdf> [Accessed 1 April 2021].

52. MINING.COM. 2021. *Ganfeng ups stake in giant Mexico lithium clay project.* [online] Available at: <https://www.mining.com/ganfeng-signs-new-jv-agreement-for-sonora/> [Accessed 6 April 2021].

53. Dubois, O. & Thiery, D. (2013). Litha and spodumene in glass. Glass International. 36. 32–34.

54. Albemarle. n.d. *Lithium | Optical Products and Glass | Albemarle.* [online] Available at: <https://www.albemarle.com/businesses/lith-ium/markets--applications/optical-products--glass#:-:text=Lithium%20Carbonate%20or%20Spodumene%20as%20Additive&text=In%20Li2CO3,%25%2C%20depending%20on%20the%20quality.> [Accessed 6 April 2021].

55. Schott.com. n.d. *SCHOTT Xensation® product variants.* [online] Available at: <https://www.schott.com/en-us/products/xensation/product-variants> [Accessed 6 April 2021].

56. Argusmedia.com. 2019. *Port Hedland raises lithium exports in September.* [online] Available at: <https://www.argusmedia.com/news/1991223-port-hedland-raises-lithium-exports-in-september> [Accessed 6 April 2021].

57. Reserve Bank of Australia. 2019. *Box B: The Recent Increase in Iron Ore Prices and Implications for the Australian Economy | Statement on Monetary Policy—August 2019.* [online] Available at: <https://www.rba.gov.au/publications/smp/2019/aug/box-b-the-recent-increase-in-iron-ore-prices-and-implications-for-the-australian-economy.html#:-:text=Australia%20is%20the%20largest%20global,per%20cent%20of%20nominal%20GDP.> [Accessed 6 April 2021].

58. Shen, M. and Zhang, M., 2019. *China considers U.S. rare earth export curbs: Global Times editor.* [online] Reuters.com Available at: <https://www.reuters.com/article/us-china-usa-rareearth-idUSKCN1SY1Gk> [Accessed 6 April 2021].

59. Ezrati, M., 2019. *China's Rare Earth Ploy.* [online] Forbes. Available at: <https://www.forbes.com/sites/miltonezrati/2019/06/14/chinas-rare-earth-ploy/?sh=6e47df777b6c> [Accessed 6 April 2021].

60. Argusmedia.com. 2020. *China's rare earth consolidation to cut supplies.* [online] Available at: <https://www.argusmedia.com/en/news/2054597-chinas-rare-earth-consolidation-to-cut-supplies> [Accessed 6 April 2021].

61. News.metal.com. 2018. *Decryption of China's four major salt lakes, five major refining technical routes! Everything about lithium extraction from the salt lake is here!_SMM | Shanghai Non ferrous Metals.* [online] Available at: <https://news.metal.com/newscontent/100911546/decryption-of-chinas-four-major-salt-lakes-five-major-refining-technical-routes-everything-about-lithium-extraction-from-the-salt-lake-is-here/> [Accessed 6 April 2021].

62. Ganfeng Lithium Co., Ltd. 2019. *Ganfeng Lithium Annual Report 2018.* [online] Available at: <https://www1.hkexnews.hk/listedco/listconews/sehk/2019/0424/ltn201904241477.pdf> [Accessed 1 April 2021].

63. Ibid.

64. Fawthrop, A., 2020. *Top six countries with the largest lithium reserves in the world.* [online] Nsenergybusiness.com. Available at: <https://www.nsenergybusiness.com/features/six-largest-lithium-reserves-world/#:~:text=The%20Greenbushes%20lithium%20mine%20in%20Western%20Australia%20%E2%80%93%20a%20joint%20venture,project%20to%20extract%20the%20metal.> [Accessed 6 April 2021].

65. Argusmedia.com. 2020. *China's Tianqi to sell Australia lithium stake to IGO.* [online] Available at: <https://www.argusmedia.com/en/news/2167267-chinas-tianqi-to-sell-australia-lithium-stake-to-igo> [Accessed 6 April 2021].

66. Australian Government, Department of Industry, Science, Energy and Resources. 2020. *Resorces and Energy Quarterly March 2020.* [online]

Available at: <https://publications.industry.gov.au/publications/resource-sandenergyquarterlymarch2020/documents/Resources-and-Energy-Quarterly-March-2020.pdf> [Accessed 6 April 2021].

67. Lee, A. and Thornhill, J., 2020. *Tianqi to Sell Stake in Lithium Mine to Ease Loan Troubles.* [online] Bloomberg.com. Available at: <https://www.bloomberg.com/news/articles/2020–12–08/tianqi-to-sell-stake-in-top-lithium-mine-to-ease-loan-troubles?sref=TtblOutp> [Accessed 6 April 2021].

68. Bo, L., 2017. *天齐锂业董事长蒋卫平的家训：天上不会掉馅饼，只有靠勤奋_四川在线.* [online] Sichuan.scol.com.cn. Available at: <https://sichuan.scol.com.cn/dwzw/201702/55817679.html> [Accessed 7 April 2021].

69. Kawase, K., 2020. *Chairman of China's Tianqi Lithium required to lend company $117m.* [online] Nikkei Asia. Available at: <https://asia.nikkei.com/Business/Markets/China-debt-crunch/Chairman-of-China-s-Tianqi-Lithium-required-to-lend-company-117m> [Accessed 7 April 2021].

70. Ibid.

71. Baike.baidu.com. n.d. *蒋卫平（天齐锂业股份公司董事长）_百度百科.* [online] Available at: <https://baike.baidu.com/item/%E8%92%8B%E5%8D%AB%E5%B9%B3/6319909?fr=aladdin> [Accessed 7 April 2021].

72. Xueqiu.com. 2019. *专访|天齐锂业董事长蒋卫平：不忘初心，坚守实业，我喜欢听机器轰隆隆的声音 天齐锂业创始人、董事长蒋卫平 锂，作为锂电产业的基础元素，被誉为是21世纪的"能源金属"和"推动世界前进的元素"。近年来… - 雪球.* [online] Available at: <https://xueqiu.com/8255210434/132620802> [Accessed 9 March 2021].

73. Sohu.com. 2019. *蒋卫平豪赌跨国并购 天齐锂业告别暴利时代_同比.* [online] Available at: <https://www.sohu.com/a/331697041_100011510> [Accessed 7 April 2021].

74. Ibid.

75. Ibid.

76. Ibid.

77. Hurun.net. 2019. *Hurun Report—Info—LEXUS Hurun China Rich List 2019.* [online] Available at: <https://www.hurun.net/en-US/Info/Detail?num=CE08472BB47d> [Accessed 7 April 2021].

78. Kawase, K., 2020. *Chairman of China's Tianqi Lithium required to lend*

company $117m. [online] Nikkei Asia. Available at: <https://asia.nikkei.com/Business/Markets/China-debt-crunch/Chairman-of-Chinas-Tianqi-Lithium-required-to-lend-company-117m> [Accessed 7 April 2021].

79. Xueqiu.com. 2019. 专访|天齐锂业董事长蒋卫平：不忘初心，坚守实业，我喜欢听机器轰隆隆的声音 天齐锂业创始人、董事长蒋卫平 锂，作为锂电产业的基础元素，被誉为是21世纪的"能源金属"和"推动世界前进的元素"。近年来...- 雪球. [online] Available at: <https://xueqiu.com/8255210434/132620802> [Accessed 9 March 2021].

80. Ingram, T., 2017. *China's Tianqi Lithium built from 'faith' in the world's lightest metal.* [online] Australian Financial Review. Available at: <https://www.afr.com/companies/mining/chinas-tianqi-lithium-built-from-faith-in-the-worlds-lightest-metal-20171019-gz4j0k> [Accessed 7 April 2021].

81. Jiaheu.com. 2015. 专访天齐锂业蒋卫平：从冷板凳到跨国并购 - 家核优居. [online] Available at: <https://www.jiaheu.com/topic/9116.html> [Accessed 7 April 2021].

82. Ibid.

83. Ibid.

84. De la Jara, A., 2018. *Tianqi buys stake in lithium miner SQM from Nutrien for $4.1 billion.* [online] Reuters.com. Available at: <https://www.reuters.com/article/us-chile-tianqi-lithium-idUSKBN1O217f> [Accessed 15 March 2021].

85. Nutrien. n.d. *Potash.* [online] Available at: <https://www.nutrien.com/what-we-do/our-business/potash#:~:text=Nutrien%20is%20the%20world's%20largest,world's%20long%2Dterm%20potash%20needs.> [Accessed 7 April 2021].

86. Jamasmie, C., 2018. *Chile antitrust watchdog probing Tianqi buy of lithium miner stake.* [online] MINING.COM. Available at: <https://www.mining.com/chile-antitrust-watchdog-probing-tianqi-buy-lithium-miner-stake/> [Accessed 7 April 2021].

87. Energymetalnews.com. 2018. *Nutrien is selling most of its stake in Chilean lithium miner for $4.07 billion.* [online] Available at: <https://energymetalnews.com/2018/05/18/nutrien-is-selling-most-of-its-stake-in-chilean-lithium-miner-for-4-07-billion/> [Accessed 7 April 2021].

88. S25.q4cdn.com. 2018. *SQM Annual Report 2018.* [online] Available at: <https://s25.q4cdn.com/757756353/files/doc_financials/2018/ar/Memoria-Anual-2018_eng.pdf> [Accessed 7 April 2021].

89. Karlsson, F., n.d. *Carey and Tianqi on the largest deal in the Chilean stock market history.* [online] Carey.cl. Available at: <https://www.carey.cl/en/carey-and-tianqi-on-the-largest-deal-in-the-chilean-stock-market-history/> [Accessed 7 April 2021].

90. Ibid.

91. Lague, D., 2005. *Obituary: Rong Yiren, 89, China's famed 'red capitalist'.* [online] Nytimes.com. Available at: <https://www.nytimes.com/2005/10/27/world/asia/obituaryrong-yiren-89-chinas-famed-red-capitalist.html> [Accessed 15 March 2021].

92. Karlsson, F., n.d. *Carey and Tianqi on the largest deal in the Chilean stock market history.* [online] Carey.cl. Available at: <https://www.carey.cl/en/carey-and-tianqi-on-the-largest-deal-in-the-chilean-stock-market-history/> [Accessed 7 April 2021].

93. Forbes. 2021. *Julio Ponce Lerou.* [online] Available at: <https://www.forbes.com/profile/julio-ponce-lerou/?sh=53e830595484> [Accessed 7 April 2021].

94. Menafn.com. 2018. *Chilean court authorizes Chinese group's lithium production purchase.* [online] Available at: <https://menafn.com/1097622039/Chilean-court-authorizes-Chinese-groups-lithium-production-purchase> [Accessed 7 April 2021].

95. Ibid.

96. Sherwood, D. and Iturrieta, F., 2018. *Exclusive: Chile files complaint to block sale of SQM shares to Chinese companies.* [online] Reuters.com. Available at: <https://www.reuters.com/article/us-chile-lithium-china-exclusive-idUSKCN1GL2Lp> [Accessed 7 April 2021].

97. BNamericas.com. 2018. *Is a cartel emerging in the global lithium market?* [online] Available at: <https://www.bnamericas.com/en/features/is-a-cartel-emerging-in-the-global-lithium-market-> [Accessed 7 April 2021].

98. Sherwood, D. and Iturrieta, F., 2018. *Exclusive: Chile files complaint to block sale of SQM shares to Chinese companies.* [online] Reuters.com. Available at: <https://www.reuters.com/article/us-chile-lithium-china-exclusive-idUSKCN1GL2Lp> [Accessed 7 April 2021].

99. Dawei, K., 2018. *Chilean Court Backs Tianqi Purchase of Minority Stake in Lithium Producer SQM.* [online] Caixinglobal.com. Available at: <https://www.caixinglobal.com/2018–10–06/chilean-court-backs-tianqi-purchase-of-minority-stake-in-lithium-producer-sqm-101 332245.html> [Accessed 7 April 2021].

100. Jamasmie, C., 2020. *China's top lithium miner struggles to pay $6 billion debt.* [online] MINING.COM. Available at: <https://www.mining.com/chinas-top-lithium-miner-struggles-to-pay-6-billion-debt/> [Accessed 7 April 2021].

101. Ibid.

102. Lin, K., Lu, X., Zhang, J. and Zheng, Y., 2020. 'State-owned enterprises in China: A review of 40-years of research and practice'. *China Journal of Accounting Research*, 13(1), pp. 31–55.

103. Ft.com. 2020. *Fall of China's 'most profitable' coal miner is a cautionary tale.* [online] Available at: <https://www.ft.com/content/f1abbb06–3f7b-469a-bca8-1996b838da2a> [Accessed 7 April 2021].

104. Reuters.com. 2020. *Australia's IGO to take 25% stake in Greenbushes lithium mine from China's Tianqi.* [online] Available at: <https://www.reuters.com/article/us-tianqi-lithium-divestiture-igo-idUSK-BN28I39g> [Accessed 7 April 2021].

105. Niewenhuis, L., 2020. *The 14 sins of Australia: Beijing expands list of grievances and digs in for extended diplomatic dispute.* [online] SupChina. Available at: <https://supchina.com/2020/11/18/the-14-sins-of-australia-beijing-expands-list-of-grievances-and-digs-in-for-extended-diplomatic-dispute/> [Accessed 7 April 2021].

106. Thomas, D. and Thomas, S., 2012. *HK bourse agrees to buy London Metal Exchange.* [online] Jp.reuters.com. Available at: <https://jp.reuters.com/article/instant-article/idINBRE85E0DU20120615> [Accessed 7 April 2021].

107. Mookerjee, I., Hu, F. and Lee, M., 2019. *Japan Companies Are Sitting on Record $4.8 Trillion in Cash.* [online] Bloomberg.com. Available at: <https://www.bloomberg.com/news/articles/2019–09–02/japan-s-companies-are-sitting-on-record-4-8-trillion-cash-pile?sref=TtblOutp> [Accessed 7 April 2021].

108. Iosebashvili, I., 2019. *Investors Look to the Yen for Dollar, Euro Insight.*

[online] Wall Street Journal. Available at: <https://www.wsj.com/articles/investors-look-to-the-yen-for-dollar-euro-insight-11577538002> [Accessed 7 April 2021].

109. Forbes. 2012. *1. Saudi Aramco—12.5 million barrels per day.* [online] Available at: <https://www.forbes.com/pictures/mef45ggld/1-saudi-aramco-12-5-million-barrels-per-day/?sh=323d5f9b6285> [Accessed 7 April 2021].

110. Riley, C., 2019. *The world has its first $2 trillion company. But for how long?.* [online] CNN. Available at: <https://edition.cnn.com/2019/12/12/investing/saudi-aramco-2-trillion/index.html> [Accessed 7 April 2021].

111. Energy.gov. n.d. *Argonne Lab's Breakthrough Cathode Technology Powers Electric Vehicles of Today.* [online] Available at: <https://www.energy.gov/articles/argonne-lab-s-breakthrough-cathode-technology-powers-electric-vehicles-today> [Accessed 7 April 2021].

112. Jaskula, B., 2021. *Lithium.* [online] Pubs.usgs.gov. Available at: <https://pubs.usgs.gov/periodicals/mcs2021/mcs2021-lithium.pdf> [Accessed 8 April 2021].

113. Sherwood, D., 2020. *Exclusive: Lithium giants Albemarle and SQM battle over access to Atacama water study.* [online] Reuters.com Available at: <https://www.reuters.com/article/us-chile-lithium-albemarle-exclusive/exclusive-lithium-giants-albemarle-and-sqm-battle-over-access-to-atacama-water-study-idUKKBN27X10g> [Accessed 6 April 2021].

114. CNBC. 2019. *White House Trade Advisor Peter Navarro Speaks with CNBC's "Squawk Box" Today.* [online] Available at: <https://www.cnbc.com/2019/09/10/cnbc-excerpts-white-house-trade-advisor-peter-navarro-speaks-with-cnbcs-squawk-box-today.html> [Accessed 8 April 2021].

115. U.S. Senate Committee on Energy and Natural Resources. 2019. *Murkowski, Manchin, Colleagues Introduce Bipartisan Legislation to Strengthen America's Mineral Security.* [online] Available at: <https://www.energy.senate.gov/2019/5/murkowski-manchin-colleagues-introduce-bipartisan> [Accessed 8 April 2021].

116. Benchmark Mineral Intelligence. 2019. *US Senator Murkowski launches*

American Mineral Security Act at Benchmark Minerals Summit in Washington DC. [online] Available at: <https://www.benchmarkminerals.com/senator-murkowski-us-government-launch-american-minerals-security-act-at-benchmark-minerals-summit-in-washington-dc/> [Accessed 8 April 2021].

117. Kane, M., 2014. *Daimler Subsidiary Li-Tec Will Cease Lithium-Ion Battery Production In December 2015*. [online] InsideEVs. Available at: <https://insideevs.com/news/323946/daimler-subsidiary-li-tec-will-cease-lithium-ion-battery-production-in-december-2015/> [Accessed 8 April 2021].
118. Loveday, S., 2016. *Daimler CEO Says There's Massive Overcapacity In Battery Cell Market*. [online] InsideEVs. Available at: <https://insideevs.com/news/328596/daimler-ceo-says-theres-massive-overcapacity-in-battery-cell-market/> [Accessed 8 April 2021].
119. Flaherty, N., 2017. *EU warns on lack of battery manufacturing in Europe*. [online] eeNews Power. Available at: <https://www.eenewspower.com/news/eu-warns-lack-battery-manufacturing-europe> [Accessed 8 April 2021].
120. Simon, F., 2017. *European battery alliance launched in Brussels*. [online] www.euractiv.com. Available at: <https://www.euractiv.com/section/electric-cars/news/european-battery-alliance-launched-in-brussels/> [Accessed 15 March 2021].
121. *Art-B, Made in Poland*. 2018. [film] Bongo Media Production, Canal+ Discovery.
122. Lambert, F., 2018. *LG is investing half a billion in its Polish battery factory to increase production*. [online] Electrek. Available at: <https://electrek.co/2018/11/30/lg-chem-polish-battery-factory-increase-production/> [Accessed 8 April 2021].
123. Northvolt.com. 2020. *Northvolt raises $600 million in equity to invest in capacity expansion, R&D and giga-scale recycling*. [online] Available at: <https://northvolt.com/newsroom/Northvolt-Sept2020> [Accessed 8 April 2021].
124. Reuters.com. 2021. *Samsung SDI to invest $849 mln to expand EV battery plant in Hungary*. [online] Available at: <https://www.reuters.com/article/samsung-sdi-batteries-hungary-idUSL4N2KU0Fs> [Accessed 8 April 2021].

125. Reuters.com. 2019. *EU to investigate Hungarian state aid for Samsung SDI's battery cell plant.* [online] Available at: <https://www.reuters.com/article/us-eu-samsungsdi-hungary-idUKKBN1WT16m> [Accessed 8 April 2021].

126. LinkedIn n.d. [online] Available at: <https://www.linkedin.com/in/ecspcar/?originalSubdomain=se> [Accessed 8 April 2021].

127. Hellstrom, J. and Pollard, N., 2021. *Sweden's Northvolt raises $1 billion to complete funding for mammoth battery plant.* [online] Reuters.com. Available at: <https://cn.reuters.com/article/us-northvolt-funding-electric-idUSKCN1TD1Wg> [Accessed 8 April 2021].

128. Reiser, A., 2019. *LG Chem battery gigafactory in Poland to be powered by EBRD.* [online] Ebrd.com. Available at: <https://www.ebrd.com/news/2019/lg-chem-battery-gigafactory-in-poland-to-be-powered-by-ebrd.html> [Accessed 8 April 2021].

129. Northvolt.com. n.d. *Production.* [online] Available at: <https://northvolt.com/production> [Accessed 8 April 2021].

130. Wattles, J., 2020. *Tesla delivered 367,500 cars last year.* [online] CNN business. Available at: <https://edition.cnn.com/2020/01/03/tech/tesla-sales/index.html> [Accessed 8 April 2021].

131. Benchmark Mineral Intelligence. 2019. *Battery megafactory capacity in the pipeline exceeds 2 TWh as solid state makes first appearance | Benchmark Mineral Intelligence.* [online] Available at: <https://www.benchmarkminerals.com/benchmarks-megafactory-tracker-exceeds-2-terawatt-hours-as-solid-state-makes-its-first-appearance/> [Accessed 8 April 2021].

132. Statista. n.d. *Car production: Number of cars produced worldwide 2018 | Statista.* [online] Available at: <https://www.statista.com/statistics/262747/worldwide-automobile-production-since-2000/> [Accessed 8 April 2021].

133. Geman, B., 2020. *Global electric vehicle sales topped 2 million in 2019.* [online] Axios. Available at: <https://www.axios.com/electric-vehicles-worldwide-sales-2fea9c70-411f-47d3-9ec6-487c7075482c.html> [Accessed 8 April 2021].

134. Loveday, S., 2016. *Daimler CEO Says There's Massive Overcapacity In Battery Cell Market.* [online] InsideEVs. Available at: <https://

insideevs.com/news/328596/daimler-ceo-says-theres-massive-over-capacity-in-battery-cell-market/> [Accessed 8 April 2021].

135. Muller, R., 2017. *Miners eye Europe's largest lithium deposit in Czech Republic.* [online] Reuters.com. Available at: <https://www.reuters.com/article/us-czech-lithium-idUSKBN18Y25x> [Accessed 8 April 2021].

136. Ibid.

137. Europeanmet.com. 2017. *PRELIMINARY FEASIBILITY STUDY CONFIRMS CINOVEC AS POTENTIALLY LOW COST LITHIUM CARBONATE PRODUCER.* [online] Available at: <https://www.europeanmet.com/wp-content/uploads/2017/03/2017019_-_EMH_Completion_of_PFS-2.pdf> [Accessed 8 April 2021].

138. Business News. n.d. *Keith Coughlan.* [online] Available at: <https://www.businessnews.com.au/Person/Keith-Coughlan> [Accessed 8 April 2021].

139. Muller, R., 2017. *Miners eye Europe's largest lithium deposit in Czech Republic.* [online] Reuters.com. Available at: <https://www.reuters.com/article/us-czech-lithium-idUSKBN18Y25x> [Accessed 8 April 2021].

140. Forbes. n.d. *Andrej Babis.* [online] Available at: <https://www.forbes.com/profile/andrej-babis/?sh=37cd75ef21ee> [Accessed 8 April 2021].

141. Svobodova, K., 2017. *Minister: ANO head Babiš politicising lithium case.* [online] Prague Monitor / Czech News in English. Available at: <https://praguemonitor.com/news/national/11/10/2017/2017–10–11-minister-ano-head-babis-politicising-lithium-case/> [Accessed 8 April 2021].

142. AP NEWS. 2018. *Czech Republic cancels lithium deal with Australian firm.* [online] Available at: <https://apnews.com/article/96ed8c593e884f148d5a87a54c8b3d7f> [Accessed 8 April 2021].

143. Argusmedia.com. 2019. *Cez looks to buy European lithium stake.* [online] Available at: <https://www.argusmedia.com/pt/news/2019346-cez-looks-to-buy-european-lithium-stake> [Accessed 8 April 2021].

144. Industry Europe. 2020. *ČEZ considering Czech lithium battery plant,*

says minister. [online] Available at: <https://industryeurope.com/sectors/transportation/%C4%8Dez%C2%A0mulling-constructing-czech-li-ion-facility-says-minister/> [Accessed 8 April 2021].

145. Stratiotis, E., 2020. *FUEL COSTS IN OCEAN SHIPPING.* [online] Morethanshipping.com. Available at: <https://www.morethanshipping.com/fuel-costs-ocean-shipping/> [Accessed 8 April 2021].

146. Smith, S., 2019. *Tianqi fires up $700m lithium game changer in Kwinana.* [online] Thewest.com.au. Available at: <https://thewest.com.au/business/mining/tianqi-fires-up-700m-lithium-game-changer-in-kwinana-ng-b881076970z> [Accessed 8 April 2021].

147. Zhang, M. and Daly, T., 2020. *China's Tianqi postpones commissioning of Australia lithium plant amid liquidity problems.* [online] Reuters.com. Available at: <https://www.reuters.com/article/tianqilithium-australia-idUSL4N2BF0De> [Accessed 8 April 2021].

148. Lambert, F., 2016. *Breakdown of raw materials in Tesla's batteries and possible bottlenecks.* [online] Electrek. Available at: <https://electrek.co/2016/11/01/breakdown-raw-materials-tesla-batteries-possible-bottleneck/> [Accessed 15 March 2021].

149. Geman, B., 2020. *Global electric vehicle sales topped 2 million in 2019.* [online] Axios. Available at: <https://www.axios.com/electric-vehicles-worldwide-sales-2fea9c70-411f-47d3-9ec6-487c7075482c.html> [Accessed 8 April 2021].

150. Barrera, P., 2020. *Tianqi Delays Lithium Plant Expansion to Focus on Steady Production.* [online] Investing News Network. Available at: <https://investingnews.com/daily/resource-investing/battery-metals-investing/lithium-investing/tianqi-delays-lithium-hydroxide-plant-expansion-focus-steady-production/> [Accessed 8 April 2021].

151. IEA. 2020. *Electric Vehicles—Analysis.* [online] Available at: <https://www.iea.org/reports/electric-vehicles> [Accessed 8 April 2021].

152. LeVine, S., 2016. *The Powerhouse: America, China and the Great Battery War.* Penguin Books, pp. 65–72.

153. Ibid.

154. Basf.com. 2016. *International Trade Commission issues final determination that Umicore infringes BASF and Argonne National Laboratory patents.* [online] Available at: <https://www.basf.com/global/en/

media/news-releases/2016/12/p-16–404.html> [Accessed 8 April 2021].

155. Basf.com. 2017. *BASF and Argonne reach resolution with Umicore over NMC patents infringement claims.* [online] Available at: <https://www.basf.com/global/en/media/news-releases/2017/04/p-17–204.html> [Accessed 8 April 2021].

156. Umicore.pl. n.d. *Nysa.* [online] Available at: <https://www.umicore.pl/en/our-sites/nysa/> [Accessed 8 April 2021].

157. The Local Germany. 2020. *German chemical giant BASF to make car battery parts near Tesla Berlin site.* [online] Available at: <https://www.thelocal.de/20200212/german-chemical-giant-basf-to-make-car-battery-parts-near-tesla-berlin-site/> [Accessed 8 April 2021].

158. ETAuto.com. 2019. *Was 2019 a year of cheer for electric vehicle industry in India?—ET Auto.* [online] Available at: <https://auto.economictimes.indiatimes.com/news/industry/was-2019-a-year-of-cheer-for-electric-vehicle-industry-in-india/72482624> [Accessed 8 April 2021].

159. van Mead, N., 2019. *22 of world's 30 most polluted cities are in India, Greenpeace says.* [online] The Guardian. Available at: <https://www.theguardian.com/cities/2019/mar/05/india-home-to-22-of-worlds-30-most-polluted-cities-greenpeace-says> [Accessed 8 April 2021].

160. Livemint.com. 2021. *India asks state refiners to review oil import contracts with Saudi.* [online] Available at: <https://www.livemint.com/industry/energy/india-asks-state-refiners-to-review-oil-import-contracts-with-saudi-11617358254052.html> [Accessed 8 April 2021].

161. Siddiqui, H., 2019. *India looks at South America's Lithium Triangle to fulfil its increasing clean energy demands.* [online] The Financial Express. Available at: <https://www.financialexpress.com/defence/india-looks-at-south-americas-lithium-triangle-to-fulfil-its-increasing-clean-energy-demands/1484417/> [Accessed 8 April 2021].

162. Ibid.

163. Jaskula, B., 2020. *Lithium.* [online] Pubs.usgs.gov. Available at: <https://pubs.usgs.gov/periodicals/mcs2020/mcs2020-lithium.pdf> [Accessed 8 April 2021].

3. LITHIUM TRIANGLE

1. Reuters.com. 2014. *Chile securities regulator levies $164 million fine in SQM probe.* [online] Available at: <https://www.reuters.com/article/us-chile-sqm-fine-idUSKBN0GX28V20140902> [Accessed 15 March 2021].

2. El Mercurio Inversiones. 2018. *Mercado apuesta a que Ponce fusionará cascadas de SQM.* [online] Available at: <https://www.elmercurio.com/Inversiones/Noticias/Acciones/2018/01/21/Mercado-apuesta-a-que-Ponce-fusionara-cascadas-de-SQM.aspx> [Accessed 9 April 2021].

3. Radio.uchile.cl. 2016. *Autor de libro de Ponce Lerou: "Empresario financió campañas políticas para protegerse".* [online] Available at: <https://radio.uchile.cl/2016/01/21/autor-de-libro-sobre-ponce-lerou-necesitaba-proteccion-y-la-consiguio-financiando-campanas-politicas/> [Accessed 9 April 2021].

4. Cofré, V., 2019. *Ponce Lerou. Pinochet—el litio—las Cascadas—las platas políticas.* Santiago, Chile: Editorial, p. 395 (Kindle edition).

5. Ibid.

6. La Tercera. 2019. *La relación de Julio Ponce Lerou y su exsuegro, "Don Augusto".* [online] Available at: <https://www.latercera.com/la-tercera-domingo/noticia/la-relacion-julio-ponce-lerou-exsuegro-don-augusto/938782/> [Accessed 9 April 2021].

7. Cofré, V., 2019. *Ponce Lerou. Pinochet—el litio—las Cascadas—las platas políticas.* Santiago, Chile: Editorial Catalonia, p. 228 (Kindle edition).

8. Ibid.

9. La Tercera. 2019. *La relación de Julio Ponce Lerou y su exsuegro, "Don Augusto.* [online] Available at: <https://www.latercera.com/la-tercera-domingo/noticia/la-relacion-julio-ponce-lerou-exsuegro-don-augusto/938782/> [Accessed 9 April 2021].

10. Archive.org. 1975. *Covert Action in Chile 1963–73.* [online] Available at: <https://archive.org/details/Covert-Action-In-Chile-1963–1973/page/n5/mode/2up> [Accessed 9 April 2021].

11. Cofré, V., 2019. *Ponce Lerou. Pinochet—el litio—las Cascadas—las platas políticas.* Santiago, Chile: Editorial Catalonia, p. 395 (Kindle edition).

12. Ibid.

13. La Tercera. 2019. *La relación de Julio Ponce Lerou y su exsuegro, "Don Augusto".* [online] Available at: <https://www.latercera.com/la-tercera-domingo/noticia/la-relacion-julio-ponce-lerou-exsuegro-don-augusto/938782/> [Accessed 9 April 2021].

14. Cofré, V., 2019. *Ponce Lerou. Pinochet—el litio—las Cascadas—las platas políticas.* Santiago, Chile: Editorial Catalonia, p. 216 (Kindle edition).

15. La Tercera. 2019. *La relación de Julio Ponce Lerou y su exsuegro, "Don Augusto".* [online] Available at: <https://www.latercera.com/la-tercera-domingo/noticia/la-relacion-julio-ponce-lerou-exsuegro-don-augusto/938782/> [Accessed 9 April 2021].

16. Ibid.

17. Ibid.

18. Cofré, V., 2019. *Ponce Lerou. Pinochet—el litio—las Cascadas—las platas políticas.* Santiago, Chile: Editorial Catalonia, p. 291 (Kindle edition).

19. Ibid.

20. Ibid.

21. Ibid.

22. La Tercera. 2019. *La relación de Julio Ponce Lerou y su exsuegro, "Don Augusto".* [online] Available at: <https://www.latercera.com/la-tercera-domingo/noticia/la-relacion-julio-ponce-lerou-exsuegro-don-augusto/938782/> [Accessed 9 April 2021].

23. Ibid.

24. Vargas Rojas, V., 2014. *¿Quién es Julio Ponce Lerou? El funcionario público yerno de Pinochet que se convirtió en millonario.* [online] El Desconcierto—Prensa digital libre. Available at: <https://www.eldesconcierto.cl/nacional/2014/09/03/quien-es-julio-ponce-lerou-el-funcionario-publico-que-se-convirtio-en-millonario.html> [Accessed 9 April 2021].

25. Cofré, V., 2019. *Ponce Lerou. Pinochet—el litio—las Cascadas—las platas políticas.* Santiago, Chile: Editorial Catalonia, p. 1906–1961 (Kindle edition).

26. Ibid.

27. Ibid.

28. Sanderson, H., 2018. *Chilean billionaire Ponce Lerou rejoins lithium producer SQM.* [online] Ft.com. Available at: <https://www.ft.com/content/225ab6a4-68e4-11e8-b6eb-4acfcfb08c11> [Accessed 9 April 2021].

29. Ibid.

30. BBC News. 2014. *Chile fines Pinochet's ex son-in-law Julio Ponce.* [online] Available at: <https://www.bbc.co.uk/news/world-latin-america-2904 2478> [Accessed 9 April 2021].

31. Urquieta, C. and Sepulveda, N., 2014. *Los grupos económicos chilenos van de tour por los paraísos fiscales.* [online] El Mostrador. Available at: <https://www.elmostrador.cl/mercados/destacados-mercado/2014/11/19/los-grupos-economicos-chilenos-van-de-tour-por-los-paraisos-fiscales/> [Accessed 9 April 2021].

32. Ibid.

33. Cofré, V., 2019. *Ponce Lerou. Pinochet—el litio—las Cascadas—las platas políticas.* Santiago, Chile: Editorial Catalonia, p. 3585–3601 (Kindle edition).

34. Ibid.

35. Reuters.com. 2013. *Chile regulator accuses SQM over market manipulation.* [online] Available at: <https://www.reuters.com/article/chile-sqm-trading-idUSL2N0H70SC20130911> [Accessed 10 April 2021].

36. Cmfchile.cl. 2014. *SVS sanciona a personas, ejecutivos y corredora de bolsa en el marco de la investigación sobre Sociedades Cascada—CMF Chile—Prensa y Presentaciones.* [online] Available at: <https://www.cmfchile.cl/portal/prensa/615/w3-article-17480.html> [Accessed 10 April 2021].

37. Vasquez, M., 2019. *Un breve análisis del caso Cascadas: buen gobierno corporativo, interés social y responsabilidades de los administradores.* [online] Elmercurio.com. Available at: <https://www.elmercurio.com/legal/movil/detalle.aspx?Id=907523&Path=/0D/D9/> [Accessed 10 April 2021].

38. CNN. 2013. *Abogado querellante detalló el caso de las empresas "cascadas".* [video] Available at: <https://www.youtube.com/watch?v=wTe O480-TtU&t=592s> [Accessed 10 April 2021].

39. Cmfchile.cl. 2014. *SVS sanciona a personas, ejecutivos y corredora de bolsa en el marco de la investigación sobre Sociedades Cascada—CMF Chile—*

Prensa y Presentaciones. [online] Available at: <https://www.cmfchile.cl/portal/prensa/615/w3-article-17480.html> [Accessed 10 April 2021].

40. Ibid.
41. Craze, M. and Quiroga, J., 2014. *SQM's Ponce Fined $70 Million in Record Chile Sanctions*. [online] Bloomberg.com. Available at: <https://www.bloomberg.com/news/articles/2014-09-02/sqm-s-ponce-fined-70-million-in-record-chile-sanctions?sref=TtblOutp> [Accessed 10 April 2021].
42. Zulver, J. and Youkee, M., 2018. *Mystical islanders divided over Chile's giant bridge project*. [online] Reuters.com. Available at: <https://www.reuters.com/article/us-chile-landrights-island-idUSKCN1GD5Je> [Accessed 16 March 2021].
43. Forbes. 2020. *Sebastian Piñera & family*. [online] Available at: <https://www.forbes.com/profile/sebastian-pinera/?sh=7d839eae7a75> [Accessed 10 March 2020].
44. Forbes. 2020. *Donald Trump*. [online] Available at: <https://www.forbes.com/profile/donald-trump/?sh=2da34caa47bd> [Accessed 10 March 2020].
45. Mitchell, C., 2018. *Is Pinochet's shadow over Chile beginning to recede?*. [online] Aljazeera.com. Available at: <https://www.aljazeera.com/news/2018/9/11/is-pinochets-shadow-over-chile-beginning-to-recede> [Accessed 16 March 2021].
46. Aravena, L., 2014. *J. Ponce: "Si su excelencia el Presidente no hubiera participado en las cascadas, no habría caso cascadas"*. [online] La Tercera. Available at: <https://www.latercera.com/noticia/j-ponce-si-su-excelencia-el-presidente-no-hubiera-participado-en-las-cascadas-no-habria-caso-cascadas/> [Accessed 10 April 2021].
47. Reuters.com. 2007. *UPDATE 3-Chile's LAN director Pinera fined over share trade*. [online] Available at: <https://www.reuters.com/article/chile-lan-pinera-idUKN0619647720070706> [Accessed 10 April 2021].
48. Sec.gov. 2017. *SEC.gov | Chemical and Mining Company in Chile Paying $30 Million to Resolve FCPA Cases*. [online] Available at: <https://www.sec.gov/news/pressrelease/2017-13.html> [Accessed 10 April 2021].
49. Sec.gov. 2018. *ORDER INSTITUTING CEASE-AND-DESIST PRO-*

CEEDINGS, PURSUANT TO SECTION 21C OF THE SECURITIES EXCHANGE ACT OF 1934, MAKING FINDINGS, AND IMPOSING REMEDIAL SANCTIONS AND A CEASE-AND-DESIST ORDER. [online] Available at: <https://www.sec.gov/litigation/admin/2018/34-84280.pdf> [Accessed 10 April 2021].

50. Cofré, V., 2016. *El mapa político de los pagos de SQM.* [online] La Tercera. Available at: <https://www.latercera.com/noticia/el-mapa-politico-de-los-pagos-de-sqm/> [Accessed 10 April 2021].

51. Americaeconomia.com. 2015. *SQM asegura que entregó de US$8,2 millones para política en Chile.* [online] Available at: <https://www.americaeconomia.com/negocios-industrias/sqm-asegura-que-entrego-de-us82-millones-para-politica-en-chile> [Accessed 10 April 2021].

52. Statista. 2020. *Copper export value in Chile 2019.* [online] Available at: <https://www.statista.com/statistics/795592/chile-value-of-copper-exports/> [Accessed 10 April 2021].

53. Reuters.com. 2019. *Chilean lithium exports continue rise, to $949 mln in 2018—central bank.* [online] Available at: <https://www.reuters.com/article/chile-lithium-idUSL1N1Z700u> [Accessed 10 April 2021].

54. Jaskula, B., 2020. *Lithium.* [online] Pubs.usgs.gov. Available at: <https://pubs.usgs.gov/periodicals/mcs2020/mcs2020-lithium.pdf> [Accessed 8 April 2021].

55. Wits.worldbank.org. n.d. *Chile Trade Summary 2018 | WITS | Text.* [online] Available at: <https://wits.worldbank.org/CountryProfile/en/Country/CHL/Year/LTST/Summarytext> [Accessed 10 April 2021].

56. Trendeconomy.com. n.d. *Saudi Arabia | Imports and Exports | World | ALL COMMODITIES | Value (US$) and Value Growth, YoY (%) | 2008–2019.* [online] Available at: <https://trendeconomy.com/data/h2/SaudiArabia/TOTAl> [Accessed 10 April 2021].

57. Miningglobal.com. 2019. *Exclusive: Wealth Minerals—"Chile can be the Saudi Arabia of lithium" | Supply Chain & Operations | Mining Global.* [online] Available at: <https://www.miningglobal.com/supply-chain-and-operations/exclusive-wealth-minerals-chile-can-be-saudi-arabia-lithium> [Accessed 10 April 2021].

58. Cubillos, C., Aguilar, P., Grágeda, M. and Dorador, C., 2018. 'Microbial Communities From the World's Largest Lithium Reserve, Salar de

Atacama, Chile: Life at High LiCl Concentrations'. *Journal of Geo-physical Research: Biogeosciences*, 123(12), pp. 3668–3681.

59. Ibid.

60. At-minerals.com. 2017. *Boom—New era for lithium—Mineral Processing*. [online] Available at: <https://www.at-minerals.com/en/artikel/at_2017–06_Boom_-_New_era_for_lithium_2849521.html> [Accessed 10 April 2021].

61. Fighter, F., 2019. *SQM: Even More Pain Ahead For Lithium*. [online] Seekingalpha.com. Available at: <https://seekingalpha.com/article/4287556-sqm-even-pain-ahead-for-lithium> [Accessed 10 April 2021].

62. Perotti, R. and Coviello, M., 2015. *GOVERNANCE OF STRATEGIC MINERALS IN LATIN AMERICA: THE CASE OF LITHIUM*. [online] Cepal.org. Available at: <https://www.cepal.org/sites/default/files/publication/files/38961/S1500861_en.pdf> [Accessed 11 April 2021].

63. Indmin.com. 2015. *What to expect from the Chilean lithium industry after the National Commission's advancement plan?* [online] Available at: <http://www.indmin.com/events/download.ashx/document/speaker/8180/a0ID000000X0kKuMAJ/Presentation> [Accessed 11 April 2021].

64. Metalbulletin.com. n.d. *Molymet—Molibdenos y Metales SA | Metal Bulletin Company Database*. [online] Available at: <https://www.metalbulletin.com/companydata/Basic-Information/Molymet-Molibdenos-y-Metales-SA/782> [Accessed 11 April 2021].

65. Encyclopedia.com. n.d. *AMAX Inc.* [online] Available at: <https://www.encyclopedia.com/books/politics-and-business-magazines/amax-inc> [Accessed 11 April 2021].

66. Perotti, R. and Coviello, M., 2015. *GOVERNANCE OF STRATEGIC MINERALS IN LATIN AMERICA: THE CASE OF LITHIUM*. [online] Cepal.org. Available at: <https://www.cepal.org/sites/default/files/publication/files/38961/S1500861_en.pdf> [Accessed 11 April 2021].

67. Cofré, V., 2019. *Ponce Lerou. Pinochet—el litio—las Cascadas—las platas políticas*. Santiago, Chile: Editorial Catalonia, p. 3346 (Kindle edition).

68. Ibid.

69. Albemarle.com. n.d. *North America | Lithium Sites & Contacts |*

Albemarle. [online] Available at: <https://www.albemarle.com/businesses/lithium/locations/north-america> [Accessed 11 April 2021].

70. Cofré, V., 2019. *Ponce Lerou. Pinochet—el litio—las Cascadas—las platas políticas*. Santiago, Chile: Editorial Catalonia, p. 3352 (Kindle edition).

71. Ibid.

72. Ibid.

73. Today's Motor Vehicles. 2016. *FMC to triple lithium hydroxide production to feed electric vehicle demand.* [online] Available at: <https://www.todaysmotorvehicles.com/article/fmc-lithium-expansion-electric-vehicles-battery-052416/> [Accessed 11 April 2021].

74. Cofré, V., 2019. *Ponce Lerou. Pinochet—el litio—las Cascadas—las platas políticas*. Santiago, Chile: Editorial Catalonia, p. 3398 (Kindle edition).

75. Ibid. p. 3411

76. Ibid.

77. Ibid.

78. Ibid., pp. 3411–3426.

79. Ibid.

80. Ibid. p. 3484.

81. Jaskula, B., 2020. *Lithium.* [online] Pubs.usgs.gov. Available at: <https://pubs.usgs.gov/periodicals/mcs2020/mcs2020-lithium.pdf> [Accessed 8 April 2021].

82. Ibid. p. 3424.

83. Sherwood, D., 2019. *Exclusive: Chile nuclear watchdog weighs probe into fraud over lithium exports—documents.* [online] Reuters.com. Available at: <https://www.reuters.com/article/us-chile-lithium-exclusive-id USKCN1P91Yg> [Accessed 11 April 2021].

84. Sanderson, H., 2016. *Lithium: Chile's buried treasure.* [online] Ft.com. Available at: <https://www.ft.com/content/cde8f984–43c7–11e6-b22f-79eb4891c97d> [Accessed 16 March 2021].

85. BNamericas.com. 2017. *SQM fails to reach agreement with Corfo.* [online] Available at: <https://www.bnamericas.com/en/news/sqm-fails-to-reach-agreement-with-corfo> [Accessed 16 March 2021].

86. BNamericas.com. 2015. *SQM fires back at Corfo over land concession*

payments. [online] Available at: <https://www.bnamericas.com/en/news/sqm-fires-back-at-corfo-over-land-concession-payments1> [Accessed 11 April 2021].

87. Arellano, A., 2018. *SQM-CORFO: las jugadas maestras que consolidaron el poder de Ponce Lerou.* [online] CIPER Chile. Available at: <https://www.ciperchile.cl/2018/06/13/sqm-corfo-las-jugadas-maestras-que-consolidaron-el-poder-de-ponce-lerou/> [Accessed 11 April 2021].

88. Reuters.com. 2016. *Chile begins new arbitration against SQM over contract dispute.* [online] Available at: <https://www.reuters.com/article/sqm-arbitration-idLTAL2N18K1F1> [Accessed 11 April 2021].

89. BNamericas.com. 2016. *SQM rejects Corfo's new arbitration request.* [online] Available at: <https://www.bnamericas.com/en/news/sqm-rejects-corfos-new-arbitration-request1> [Accessed 11 April 2021].

90. BNamericas.com. 2016. *Corfo launches 2nd arbitration process against SQM.* [online] Available at: <https://www.bnamericas.com/en/news/corfo-launches-2nd-arbitration-process-against-sqm1> [Accessed 11 April 2021].

91. Ibid.

92. BNamericas.com. 2016. *SQM hits back in dispute with state agency.* [online] Available at: <https://www.bnamericas.com/en/news/sqm-hits-back-in-dispute-with-state-agency/?position=687864> [Accessed 11 April 2021].

93. Roskill.com. 2018. *Lithium: Corfo and SQM settle differences, agree new Salar de Atacama license.* [online] Available at: <https://roskill.com/news/lithium-corfo-sqm-settle-differences-agree-new-salar-de-atacama-license/> [Accessed 11 April 2021].

94. Ibid.

95. Lowry, J., Hersch, E. and Bitran, E., 2018. *Episode 27 "Man on a Mission".* [podcast] Global Lithium Podcast. Available at: <http://lithiumpodcast.com/podcast/e27-man-on-a-mission/> [Accessed 15 March 2021].

96. Ibid.

97. Jamasmie, C., 2019. *EV sector will need 250% more copper by 2030 just for charging stations.* [online] MINING.COM. Available at: <https://

www.mining.com/ev-sector-will-need-250-more-copper-by-2030-just-for-charging-stations/#:~:text=While%20it%20is%20 a%20known,growth%2C%20a%20new%20study%20shows.> [Accessed 11 April 2021].

98. Lowry, J., Hersh, E., Galli, D., Galli, C. and Alvarado, D., 2018. *Episode 11: Lithium Family Values in Argentina.* [podcast] Global Lithium Podcast. Available at: <http://lithiumpodcast.com/podcast/ e11-lithium-family-values-in-argentina/> [Accessed 15 March 2021].

99. Ibid.

100. Ibid.

101. Guzman, J., Faundez, P., Jara, J. and Retamal, C., 2021. ROLE OF LITHIUM MINING ON THE WATER STRESS OF THE SALAR DE ATACAMA BASIN.

102. Lowry, J., Hersh, E., Galli, D., Galli, C. and Alvarado, D., 2018. *Episode 11: Lithium Family Values in Argentina.* [podcast] Global Lithium Podcast. Available at: <http://lithiumpodcast.com/podcast/ e11-lithium-family-values-in-argentina/> [Accessed 15 March 2021].

103. Ibid.

104. Ibid.

105. Ibid.

106. Ibid.

107. Ibid.

108. Ibid.

109. Ibid.

110. Iturrieta, F. and O'Brien, R., 2017. *Chile to invite bids on value-added lithium tech in April.* [online] Reuters.com. Available at: <https:// www.reuters.com/article/us-chile-corfo-idUSKBN15127u> [Accessed 11 April 2021].

111. Lombrana, L., 2019. *Lithium-Rich Chile Seeks to Become Major Player in Battery Sector.* [online] Transport Topics. Available at: <https:// www.ttnews.com/articles/lithium-rich-chile-seeks-become-major-player-battery-sector> [Accessed 11 April 2021].

112. Ibid.

113. Lombrana, L., 2019. *Lithium Giant's Landmark Deal Is Big Step to Battery 'Dream'.* [online] Bloomberg.com. Available at: <https://www.

bloomberg.com/news/articles/2019–02–01/lithium-giant-s-landmark-deal-is-big-step-to-battery-hub-dream?sref=TtblOutp> [Accessed 11 April 2021].

114. Sherwood, D., 2019. *Molymet drops plans for battery parts factory in Chile.* [online] Reuters.com. Available at: <https://www.reuters.com/article/chile-lithium/molymet-drops-plans-for-battery-parts-factory-in-chile-idUKL2N24E04t> [Accessed 11 April 2021].

115. Chung, J. and Sherwood, D., 2019. South Korea's POSCO drops plans for Chilean battery material plant. [online] Reuters.com. Available at: <https://www.reuters.com/article/us-chile-lithium-posco-idUSK CN1TM2Lr> [Accessed 11 April 2021].

116. Sherwood, D., 2019. *RPT-FOCUS-How lithium-rich Chile botched a plan to attract battery makers.* [online] Reuters.com. Available at: <https://www.reuters.com/article/chile-lithium-idUSL2N24H1W8> [Accessed 11 April 2021].

117. Cofré, V., 2019. *Ponce Lerou. Pinochet—el litio—las Cascadas—las platas políticas.* Santiago, Chile: Editorial Catalonia, p. 3475–3488 (Kindle edition).

118. Ibid.

119. Ibid.

120. Ibid. p. 3502.

121. Ibid.

122. Ibid. p. 3516.

123. Ober, J., 1995. *Lithium.* [online] S3-us-west-2.amazonaws.com. Available at: <https://s3-us-west-2.amazonaws.com/prd-wret/assets/palladium/production/mineral-pubs/lithium/450495.pdf> [Accessed 15 March 2021].

124. Maxwell, P., 2013. Analysing the lithium industry: Demand, supply, and emerging developments. *Mineral Economics*, 26(3), pp. 97–106.

125. Cofré, V., 2019. *Ponce Lerou. Pinochet—el litio—las Cascadas—las platas políticas.* Santiago, Chile: Editorial Catalonia, p. 3516

126. Ibid. pp. 3516–3544.

127. Lowry, J., Hersch, E. and Bitran, E., 2018. *Episode 27 "Man on a Mission".* [podcast] Global Lithium Podcast. Available at: <http://lithiumpodcast.com/podcast/e27-man-on-a-mission/> [Accessed 15 March 2021].

128. Ibid.
129. Ibid.
130. Ibid.
131. Ibid.
132. Ibid.
133. Ibid.
134. Yáñez, D., 2016. *De Corfo a SQM: El evidente conflicto de interés de Rafael Guilisasti*. [online] El Ciudadano. Available at: <https://www.elciudadano.com/economia/de-corfo-a-sqm-el-evidente-conflicto-de-interes-de-rafael-guilisasti/09/16/> [Accessed 11 April 2021].
135. Lowry, J., Hersch, E. and Bitran, E., 2018. *Episode 27 "Man on a Mission"*. [podcast] Global Lithium Podcast. Available at: <http://lithiumpodcast.com/podcast/e27-man-on-a-mission/> [Accessed 15 March 2021].
136. Ibid.
137. Ibid.
138. Ibid.
139. Steinbild, M., 2018. *SQM reached Agreement with CORFO: steinbildconsulting.com*. [online] Steinbildconsulting.com. Available at: <http://www.steinbildconsulting.com/index.php/blog/sqm-reached-agreement-corfo> [Accessed 11 April 2021].
140. Lowry, J., Hersch, E. and Bitran, E., 2018. *Episode 27 "Man on a Mission"*. [podcast] Global Lithium Podcast. Available at: <http://lithiumpodcast.com/podcast/e27-man-on-a-mission/> [Accessed 15 March 2021].
141. Ibid.
142. Ibid.
143. Ibid.
144. Ibid.
145. SQM. n.d. *Our History*. [online] Available at: <https://www.sqm.com/en/acerca-de-sqm/informacion-corporativa/nuestra-historia/> [Accessed 12 April 2021].
146. Jaskula, B., 2020. *Lithium*. [online] Pubs.usgs.gov. Available at: <https://pubs.usgs.gov/periodicals/mcs2020/mcs2020-lithium.pdf> [Accessed 8 April 2021].

147. Ibid.
148. Argusmedia.com. 2020. *Argentina hints at incentives for lithium investment.* [online] Available at: <https://www.argusmedia.com/en/news/2160377-argentina-hints-at-incentives-for-lithium-investment#:~:text=Argentina%20has%20the%20world's% 20third, undergoing% 20a% 20preliminary% 20economic% 20assessment.> [Accessed 12 April 2021].
149. Lowry, J., Hersh, E., Galli, D., Galli, C. and Alvarado, D., 2018. *Episode 11: Lithium Family Values in Argentina.* [podcast] Global Lithium Podcast. Available at: <http://lithiumpodcast.com/podcast/e11-lithium-family-values-in-argentina/> [Accessed 15 March 2021].
150. Ibid.
151. Talens Peiró, L., Villalba Méndez, G. and Ayres, R., 2013. 'Lithium: Sources, Production, Uses, and Recovery Outlook'. *JOM*, 65 (8), pp. 986–996.
152. Patents.google.com. 2013. *Process for producing lithium carbonate from concentrated lithium brine.* [online] Available at: <https://patents.google.com/patent/WO2013036983A1/en> [Accessed 12 April 2021].
153. Talens Peiró, L., Villalba Méndez, G. and Ayres, R., 2013. 'Lithium: Sources, Production, Uses, and Recovery Outlook'. *JOM*, 65 (8), pp. 986–996.
154. Peyrille, A., 2015. *Argentina's Mine Industry Doubles Down on Lithium.* [online] IndustryWeek. Available at: <https://www.industryweek.com/the-economy/environment/article/21966108/argentinas-mine-industry-doubles-down-on-lithium> [Accessed 12 April 2021].
155. Lambert, F., 2017. *Tesla officials visit Argentina's Gorvernor of Salta for solar + storage projects and sourcing lithium.* [online] Electrek. Available at: <https://electrek.co/2017/05/04/tesla-argentina-solar-storage-lithium/> [Accessed 12 April 2021].
156. BNamericas.com. 2019. *Alberto Fernández meets miners to ease investment fears.* [online] Available at: <https://www.bnamericas.com/en/features/alberto-fernandez-meets-miners-to-ease-investment-fears> [Accessed 12 April 2021].
157. Eramet.com. 2012. *Transforming much more than ore.* [online] Available at: <https://www.eramet.com/sites/default/files/2019–05/eramet_reference_document_2012.pdf> [Accessed 12 April 2021].

158. Eramet.com. 2020. *Eramet: a leader in battery recycling in Europe?* [online] Available at: <https://www.eramet.com/en/eramet-leader-battery-recycling-europe> [Accessed 12 April 2021].

159. BNamericas.com. 2020. *Eramet abandons US$600mn Argentina lithium project.* [online] Available at: <https://www.bnamericas.com/en/news/eramet-abandons-us600mn-argentina-lithium-project> [Accessed 12 April 2021].

160. Reuters.com. 2020. *France's Eramet freezes lithium mine project in Argentina.* [online] Available at: <https://www.reuters.com/article/us-eramet-results-idUSKBN20D2Kp> [Accessed 12 April 2021].

161. Marchegiani, P., Höglund, J., Gómez, H. and Gómez, L., 2019. *Lithium extraction in Argentina: a case study on the social and environmental impacts.* [online] Goodelectronics.org. Available at: <https://goodelectronics.org/wp-content/uploads/sites/3/2019/05/DOC_LITHIUM_ENGLISH.pdf> [Accessed 12 April 2021].

162. Informacionminera.produccion.gob.ar. 2019. *Argentina Advanced Lithium Projects in Salares.* [online] Available at: <http://informacionminera.produccion.gob.ar/assets/datasets/2019–07–15%20Proyectos%20Avanzados%20de%20Litio%20en%20Argentina.pdf> [Accessed 12 April 2021].

163. Srk.com. n.d. *South America: Salar De Uyuni Brine Deposit.* [online] Available at: <https://www.srk.com/en/publications/south-america-salar-de-uyuni-brine-deposit> [Accessed 12 April 2021].

164. Katwala, A., 2018. *The spiralling environmental cost of our lithium battery addiction.* [online] WIRED UK. Available at: <https://www.wired.co.uk/article/lithium-batteries-environment-impact> [Accessed 16 March 2021].

165. Obbekær, M. and Mortensen, N., 2019. *How much water is used to make the world's batteries?* [online] Danwatch.dk. Available at: <https://danwatch.dk/en/undersoegelse/how-much-water-is-used-to-make-the-worlds-batteries/> [Accessed 12 April 2021].

166. Marchegiani, P., Höglund, J., Gómez, H. and Gómez, L., 2019. *Lithium extraction in Argentina: a case study on the social and environmental impacts.* [online] Goodelectronics.org. Available at: <https://goodelectronics.org/wp-content/uploads/sites/3/2019/05/DOC_LITHIUM_ENGLISH.pdf> [Accessed 12 April 2021].

167. Ibid.
168. Un.org. n.d. *United Nations Declaration on the Rights of Indigenous Peoples.* [online] Available at: <https://www.un.org/development/desa/indigenouspeoples/wp-content/uploads/sites/19/2019/01/UNDRIP_E_web.pdf> [Accessed 12 April 2021].
169. Marchegiani, P., Höglund, J., Gómez, H. and Gómez, L., 2019. *Lithium extraction in Argentina: a case study on the social and environmental impacts.* [online] Goodelectronics.org. Available at: <https://goodelectronics.org/wp-content/uploads/sites/3/2019/05/DOC_LITHIUM_ENGLISH.pdf> [Accessed 12 April 2021].
170. Ibid.
171. Ft.com. 2016. *Lunch with the FT: Ali al-Naimi on two decades as Saudi's oil king.* [online] Available at: <https://www.ft.com/content/348ce86c-ac19-11e6-ba7d-76378e4fef24> [Accessed 15 March 2021].

4. SAUDI ARABIA OF LITHIUM

1. Refworld. 2012. *State of the World's Minorities and Indigenous Peoples 2012—Bolivia.* [online] Available at: <https://www.refworld.org/docid/4fedb407c.html> [Accessed 12 April 2021].
2. Dube, J., 2019. *Bolivian President Resigns After Re-Election Marred by Fraud Allegations.* [online] Wall Street Journal. Available at: <https://www.wsj.com/articles/bolivia-s-president-evo-morales-calls-for-new-presidential-elections-11573391449> [Accessed 12 April 2021].
3. Londoño, E., 2019. *Bolivian Leader Evo Morales Steps Down (Published 2019).* [online] Nytimes.com. Available at: <https://www.nytimes.com/2019/11/10/world/americas/evo-morales-bolivia.html> [Accessed 12 April 2021].
4. France 24. 2019. *Morales claims US orchestrated 'coup' to tap Bolivia's lithium.* [online] Available at: <https://www.france24.com/en/20191224-morales-claims-us-orchestrated-coup-to-tap-bolivia-s-lithium> [Accessed 12 April 2021].
5. Ibid.
6. Jaskula, B., 2020. [online] Pubs.usgs.gov. Available at: <https://pubs.usgs.gov/periodicals/mcs2020/mcs2020-lithium.pdf> [Accessed 18 March 2021].

7. Lithium Today. 2018. *Lithium supply in Bolivia.* [online] Available at: <http://lithium.today/lithium-supply-by-countries/lithium-supply-bolivia/> [Accessed 12 April 2021].

8. Greenfield, P., 2016. *Story of cities #6: how silver turned Potosí into 'the first city of capitalism'.* [online] The Guardian. Available at: <https://www.theguardian.com/cities/2016/mar/21/story-of-cities-6-potosi-bolivia-peru-inca-first-city-capitalism> [Accessed 12 April 2021].

9. Economist.com. 1999. *Tin Soldiers.* [online] Available at: <https://www.economist.com/the-americas/1999/01/07/tin-soldiers> [Accessed 12 April 2021].

10. Engdahl, F., 2009. *Russia and Bolivia to Launch Gas Joint Venture.* [online] Archive.globalpolicy.org. Available at: <https://archive.globalpolicy.org/challenges-to-the-us-empire/the-rise-of-competitors/48328-russia-and-bolivia-to-launch-gas-joint-venture. html> [Accessed 12 April 2021].

11. Fuentes, F., 2013. *Nationalization puts wealth in hands of the Bolivian people.* [online] Canadiandimension.com. Available at: <https://canadiandimension.com/articles/view/nationalisation-puts-wealth-in-hands-of-the-bolivian-people> [Accessed 12 April 2021].

12. Abelvik-Lawson, H., 2019. 'Indigenous Environmental Rights, Participation and Lithium Mining in Argentina and Bolivia: A Socio-Legal Analysis'. *PhD thesis, University of Essex,* [online] Available at: <http://repository.essex.ac.uk/25797/1/Helle%20A-L%20THESIS%20FINAL.pdf> [Accessed 18 March 2021].

13. Wikileaks.org. 2009. *Cable: 09LAPAZ267_a.* [online] Available at: <https://wikileaks.org/plusd/cables/09LAPAZ267_a.html> [Accessed 18 March 2021].

14. Ibid.

15. Ibid.

16. Ibid.

17. Abelvik-Lawson, H., 2019. 'Indigenous Environmental Rights, Participation and Lithium Mining in Argentina and Bolivia: A Socio-Legal Analysis'. *PhD thesis, University of Essex,* [online] Available at: <http://repository.essex.ac.uk/25797/1/Helle%20A-L%20THESIS%20FINAL.pdf> [Accessed 18 March 2021].

18. Lithium Today. 2018. *With experts on lithium (series)—JUAN CARLOS ZULETA.* [online] Available at: <http://lithium.today/expert-lithium-series/> [Accessed 18 March 2021].

19. Ibid.

20. Abelvik-Lawson, H., 2019. 'Indigenous Environmental Rights, Participation and Lithium Mining in Argentina and Bolivia: A Socio-Legal Analysis'. *PhD thesis, University of Essex,* [online] Available at: <http://repository.essex.ac.uk/25797/1/Helle%20A-L%20THESIS%20FINAL.pdf> [Accessed 18 March 2021].

21. Ibid.

22. Vidal, J., 2009. *Evo Morales stuns Copenhagen with demand to limit temperature rise to 1C.* [online] The Guardian. Available at: <https://www.theguardian.com/environment/2009/dec/16/evo-morales-hugo-chavez> [Accessed 12 April 2021].

23. Kurmanaev, A. and Krauss, C., 2019. *Ethnic Rifts in Bolivia Burst Into View With Fall of Evo Morales (Published 2019).* [online] Nytimes.com. Available at: <https://www.nytimes.com/2019/11/15/world/americas/morales-bolivia-Indigenous-racism.html> [Accessed 12 April 2021].

24. Radhuber, I. and Andreucci, D., 2014. *Indigenous Bolivians are seething over mining reforms.* [online] The Conversation. Available at: <https://theconversation.com/indigenous-bolivians-are-seething-over-mining-reforms-27085> [Accessed 12 April 2021].

25. Telesurenglish.net. 2019. *Bolivia to Introduce First Domestically-Made Electric Vehicle.* [online] Available at: <https://www.telesurenglish.net/news/Bolivia-to-Introduce-First-Domestically-Made-Electric-Vehicle--20191002–0015.html> [Accessed 12 April 2021].

26. Garcia, E., 2009. *RPT-INTERVIEW-LG may seek to tap Bolivian lithium deposit.* [online] Reuters.com. Available at: <https://www.reuters.com/article/bolivia-lithium/rpt-interview-lg-may-seek-to-tap-bolivian-lithium-deposit-idUKN0954338620090209> [Accessed 12 April 2021].

27. Rfi.fr. 2009. *Be partners not predators, Morales warns French firms on Paris visit.* [online] Available at: <http://www1.rfi.fr/actuen/articles/110/article_2926.asp> [Accessed 12 April 2021].

28. Wikileaks.org. 2009. *Cable: 09LAPAZ267_a.* [online] Available at:

<https://wikileaks.org/plusd/cables/09LAPAZ267_a.html> [Accessed 18 March 2021].

29. Ibid.

30. Metalbulletin.com. 2009. *Bolloré-Eramet lithium development partnership with Bolivia.* [online] Available at: <https://www.metalbulletin.com/Article/2164778/Bollor-Eramet-lithium-development-partnership-with-Bolivia.html> [Accessed 12 April 2021].

31. Francois, I., 2010. *Bolloré et Eramet vont chercher du lithium in Argentine.* [online] Les Echos. Available at: <https://www.lesechos.fr/2010/02/bollore-et-eramet-vont-chercher-du-lithium-en-argentine-418145> [Accessed 12 April 2021].

32. Wikileaks.org. 2009. *Cable: 09LAPAZ267_a.* [online] Available at: <https://wikileaks.org/plusd/cables/09LAPAZ267_a.html> [Accessed 18 March 2021].

33. Ibid.

34. Ibid.

35. Ibid.

36. Ibid.

37. Obayashi, Y., 2019. *Japan's SMM aims to double battery material capacity in nine years.* [online] Reuters.com. Available at: <https://www.reuters.com/article/us-sumitomo-mtl-min-metals-idUSKCN1SN0J6> [Accessed 12 April 2021].

38. Romero, S., 2009. *In Bolivia, Untapped Bounty Meets Nationalism (Published 2009).* [online] Nytimes.com. Available at: <https://www.nytimes.com/2009/02/03/world/americas/03lithium.html> [Accessed 12 April 2021].

39. Abelvik-Lawson, H., 2019. 'Indigenous Environmental Rights, Participation and Lithium Mining in Argentina and Bolivia: A Socio-Legal Analysis'. *PhD thesis, University of Essex,* [online] Available at: <http://repository.essex.ac.uk/25797/1/Helle%20A-L%20THESIS%20FINAL.pdf> [Accessed 18 March 2021].

40. Ibid.

41. Roy, F., 2016. *Bolivia's Comibol to commence lithium output from Salar de Uyuni deposit in Q4'18.* [online] Spglobal.com. Available at: <https://www.spglobal.com/marketintelligence/en/news-insights/trending/llq-ytkmjieymrx0ipcg1q2> [Accessed 12 April 2021].

42. Ylb.gob.bo. 2020. *Yacimientos de Litio Bolivianos, RENDICIÓN DE CUENTAS PÚBLICAS, 2019–2020.* [online] Available at: <https://www.ylb.gob.bo/archivos/notas_archivos/rendicion_de_cuentas_publicas_c.pdf> [Accessed 12 April 2021].

43. Aljazeera.com. 2017. *The changing landscape of Bolivia's salt flats.* [online] Available at: <https://www.aljazeera.com/gallery/2017/5/3/the-changing-landscape-of-bolivias-salt-flats> [Accessed 12 April 2021].

44. McCrae, M., 2015. *Orocobre's lithium plant is up and running.* [online] MINING.COM. Available at: <https://www.mining.com/orocobres-lithium-plant-is-up-and-running-32706/> [Accessed 12 April 2021].

45. Ingram, T., 2017. *Olaroz: Orocobre's high-altitude lithium challenge.* [online] Australian Financial Review. Available at: <https://www.afr.com/companies/mining/olaroz-orocobres-highaltitude-lithium-challenge-20171110-gzigx4> [Accessed 12 April 2021].

46. ECM. n.d. *PV equipment manufacturer Industrial PV Furnaces.* [online] Available at: <https://ecm-greentech.fr/> [Accessed 12 April 2021].

47. Mitra Taj, M., 2019. *In the new lithium 'Great Game,' Germany edges out China in Bolivia.* [online] Reuters.com. Available at: <https://www.reuters.com/article/bolivia-lithium-germany-idINKCN1PM1Q3> [Accessed 12 April 2021].

48. Ibid.

49. Lawyer Monthly. 2019. *ACISA create joint venture to industrialize lithium.* [online] Available at: <https://www.lawyer-monthly.com/2019/03/acisa-create-joint-venture-to-industrialize-lithium/> [Accessed 12 April 2021].

50. Ibid.

51. Ramos, D., 2019. *Bolivia picks Chinese partner for $2.3 billion lithium projects.* [online] Reuters.com. Available at: <https://www.reuters.com/article/us-bolivia-lithium-china-idUSKCN1PV2F7> [Accessed 12 April 2021].

52. Reuters.com. 2019. *Bolivia's lithium partnership with Germany's ACI Systems hits snag.* [online] Available at: <https://www.reuters.com/article/us-bolivia-germany-lithium-idUSKBN1XE01n> [Accessed 12 April 2021].

53. Preuss, S., 2019. *So schnell geben die Schwaben das Lithium-Projekt nicht*

verloren. [online] FAZ.NET. Available at: <https://www.faz.net/aktu­ell/wirtschaft/unternehmen/so-schnell-gibt-aci-das-lithium-projekt-nicht-verloren-16470092.html> [Accessed 12 April 2021].

54. DW.COM. 2019. *Bolivians protest over lithium deal with German company.* [online] Available at: <https://www.dw.com/en/bolivians-pro­test-over-lithium-deal-with-german-company/a-50732216> [Accessed 12 April 2021].

55. Preuss, S., 2019. *So schnell geben die Schwaben das Lithium-Projekt nicht verloren.* [online] FAZ.NET. Available at: <https://www.faz.net/aktu­ell/wirtschaft/unternehmen/so-schnell-gibt-aci-das-lithium-projekt-nicht-verloren-16470092.html> [Accessed 12 April 2021].

56. Belghaus, N. and Franke, F., 2020. *Lithiumgewinnung in Bolivien: Alles auf Weiß.* [online] Taz.de. Available at: <https://taz.de/Lithium­gewinnung-in-Bolivien/!5709257/> [Accessed 13 April 2021].

57. M.eldiario.net. 2020. *Detener proyecto del litio sería "duro revés" para relaciones.* [online] Available at: <https://www.eldiario.net/movil/index. php?n=23&a=2020&m=01&d=23#closem> [Accessed 13 April 2021].

58. Nienaber, M., 2020. *Germany to urge next Bolivian leaders to revive lithium deal.* [online] Reuters.com. Available at: <https://www.reuters. com/article/us-germany-bolivia-lithium-idUSKBN1ZM1Ip> [Accessed 13 April 2021].

59. Jemio, M., 2020. *Bolivia rethinks how to industrialise its lithium amid political transition.* [online] Dialogo Chino. Available at: <https://dia­logochino.net/en/extractive-industries/35423-bolivia-rethinks-how-to-industrialize-its-lithium-amid-political-transition/> [Accessed 13 April 2021].

60. Erbol. 2020. *Uyuni: Reciben con bloqueo a nuevo gerente de YLT Juan Carlos Zuleta.* [online] Available at: <https://erbol.com.bo/nacional/uyuni-reciben-con-bloqueo-nuevo-gerente-de-ylt-juan-carlos-zuleta> [Accessed 18 March 2021].

61. Blair, L. and Bercerra, C., 2020. *Is Bolivia's 'interim' president using the pandemic to outstay her welcome?.* [online] The Guardian. Available at: <https://www.theguardian.com/global-development/2020/jun/01/bolivia-president-jeanine-anez-coronavirus-elections> [Accessed 18 March 2021].

62. *Yacimientos de Litio Bolivianos Corporación. 2021. AUDIENCIA PÚBLICA DE RENDICIÓN DE CUENTAS FINAL 2020.* [online] Available at: <https://www.ylb.gob.bo/resources/rendicion_cuentas/audiencia_publica_2020.pdf> [Accessed 13 April 2021].

63. Jemio, M., 2020. *Bolivia rethinks how to industrialise its lithium amid political transition.* [online] Dialogo Chino. Available at: <https://dialogochino.net/en/extractive-industries/35423-bolivia-rethinks-how-to-industrialize-its-lithium-amid-political-transition/> [Accessed 13 April 2021].

64. Benchmark Mineral Intelligence. 2020. *Bolivia presidential candidate Luis Arce outlines Lithium First Industrial Strategy; Benchmark advising on commercial strategy.* [online] Available at: <https://www.benchmarkminerals.com/membership/bolivia-presidential-candidate-luis-arce-outlines-lithium-first-industrial-strategy-benchmark-advising-on-commercial-strategy/> [Accessed 13 April 2021].

65. Jemio, M., 2020. *Bolivia rethinks how to industrialise its lithium amid political transition.* [online] Dialogo Chino. Available at: <https://dialogochino.net/en/extractive-industries/35423-bolivia-rethinks-how-to-industrialize-its-lithium-amid-political-transition/> [Accessed 13 April 2021].

5. ARE WE REALLY MAKING THE WORLD A BETTER PLACE?

1. Lezhnev, S., 2016. *Why You Can't Call Congo a Failed State.* [online] Time. Available at: <https://time.com/4545223/why-you-cant-call-congo-a-failed-state/> [Accessed 13 April 2021].

2. Reuters.com. 2018. *Congo declares cobalt 'strategic', nearly tripling royalty rate.* [online] Available at: <https://www.reuters.com/article/us-congo-cobalt-idUSKBN1O220d> [Accessed 13 April 2021].

3. Shengo, M., Kime, M., Mambwe, M. and Nyembo, T., 2019. 'A review of the beneficiation of copper-cobalt-bearing minerals in the Democratic Republic of Congo'. *Journal of Sustainable Mining*, 18(4), pp. 226–246.

4. Comtrade.un.org. 2021. *Download trade data | UN Comtrade: International Trade Statistics.* [online] Available at: <https://comtrade.un.org/data> [Accessed 13 April 2021].

5. Ibid.

6. Argusmedia.com. 2020. *Cobalt giants back changes to DRC artisanal mining.* [online] Available at: <https://www.argusmedia.com/en/news/2135154-cobalt-giants-back-changes-to-drc-artisanal-mining> [Accessed 13 April 2021].

7. Amnesty.org. 2016. *"THIS IS WHAT WE DIE FOR" HUMAN RIGHTS ABUSES IN THE DEMOCRATIC REPUBLIC OF THE CONGO POWER THE GLOBAL TRADE IN COBALT.* [online] Available at: <https://www.amnesty.org/download/Documents/AFR6231832016ENGLISH.pdf> [Accessed 18 March 2021].

8. Global Witness. 2019. *Why cutting off artisanal miners is not responsible sourcing.* [online] Available at: <https://www.globalwitness.org/en/blog/why-cutting-artisanal-miners-not-responsible-sourcing/> [Accessed 13 April 2021].

9. Amnesty.org. 2016. *"THIS IS WHAT WE DIE FOR" HUMAN RIGHTS ABUSES IN THE DEMOCRATIC REPUBLIC OF THE CONGO POWER THE GLOBAL TRADE IN COBALT.* [online] Available at: <https://www.amnesty.org/download/Documents/AFR6231832016ENGLISH.pdf> [Accessed 18 March 2021].

10. Interbrand. n.d. *Four decades of growth for a global leader.* [online] Available at: <https://interbrand.com/work/bmw-2/> [Accessed 13 April 2021].

11. MINING.COM. 2020. *Cobalt price: BMW avoids the Congo conundrum—for now.* [online] Available at: <https://www.mining.com/cobalt-price-bmw-avoids-the-congo-conundrum-for-now/> [Accessed 18 March 2021].

12. Shedd, K., 2020. *Cobalt Data Sheet—Mineral Commodity Summaries 2020.* [online] Pubs.usgs.gov. Available at: <https://pubs.usgs.gov/periodicals/mcs2020/mcs2020-cobalt.pdf> [Accessed 19 March 2021].

13. France 24. 2019. *Fall in cobalt price pushes DR Congo to reform economy.* [online] Available at: <https://www.france24.com/en/20190616-fall-cobalt-price-pushes-dr-congo-reform-economy> [Accessed 13 April 2021].

14. Kisangani, N., 2000. 'Congo (Zaire): Corruption, Disintegration, and State Failure'. E. Wayne Nafziger, Frances Stewart and Raimo Väyrynen (eds), *War, Hunger, and Displacement, Volume 2.* Oxford, UK: Oxford University Press, pp. 261–294.

15. The Guardian 2001. *Revealed: how Africa's dictator died at the hands of his boy soldiers.* [online] Available at: <https://www.theguardian.com/world/2001/feb/11/theobserver> [Accessed 13 April 2021].

16. Wild, F., Kavanagh, M. and Clowes, W., 2020. *Sanctioned Billionaire Finds a Haven in Tiny Congolese Bank.* [online] Bloomberg.com. Available at: <https://www.bloomberg.com/news/features/2020-07-02/sanctioned-billionaire-dan-gertler-s-haven-a-tiny-congolese-bank?sref=TtblOutp> [Accessed 18 March 2021].

17. BBC News. n.d. *Dan Gertler—the man at the centre of DR Congo corruption allegations.* [online] Available at: <https://www.bbc.com/news/world-africa-56444576> [Accessed 13 April 2021].

18. Wilson, T., 2018. *DRC president Joseph Kabila defends Glencore and former partner Gertler.* [online] Ft.com. Available at: <https://www.ft.com/content/8c9a416a-fc6e-11e8-aebf-99e208d3e521> [Accessed 13 April 2021].

19. ReliefWeb. 2001. *Addendum to the report of the Panel of Experts on the Illegal Exploitation of Natural Resources and Other Forms of Wealth of DR Congo (S/2001/1072)—Angola.* [online] Available at: <https://reliefweb.int/report/angola/addendum-report-panel-experts-illegal-exploitation-natural-resources-and-other-forms> [Accessed 13 April 2021].

20. U.S. Department of the Treasury. 2018. *Treasury Sanctions Fourteen Entities Affiliated with Corrupt Businessman Dan Gertler Under Global Magnitsky.* [online] Available at: <https://home.treasury.gov/news/press-releases/sm0417> [Accessed 13 April 2021].

21. Wallis, W., Binham, C. and Sakoui, A., 2013. *Annan report blasts ENRC for costing Congo $725m.* [online] Ft.com. Available at: <https://www.ft.com/content/e486f064-b8c0-11e2-869f-00144feabdc0?_i_location=http%3A%2F%2Fwww.ft.com%2Fcms%2Fs%2F0%2Fe486f064-b8c0-11e2-869f-00144feabdc0.html%3Fsiteedition%3Duk&_i_referer=&siteedition=uk> [Accessed 13 April 2021].

22. Burgis, T., 2017. *Why Glencore bought Israeli tycoon out of Congo mines.* [online] Ft.com. Available at: <https://www.ft.com/content/8c4de26e-0366-11e7-ace0-1ce02ef0def9> [Accessed 13 April 2021].

23. Glencore. 2017. *Glencore purchases stakes in Mutanda and Katanga.*

[online] Available at: <https://www.glencore.com/media-and-insights/news/glencore-purchases-stakes-in-mutanda-and-katanga> [Accessed 13 April 2021].

24. U.S. Department of the Treasury. 2017. *United States Sanctions Human Rights Abusers and Corrupt Actors Across the Globe.* [online] Available at: <https://home.treasury.gov/news/press-releases/sm0243> [Accessed 13 April 2021].

25. Statista. n.d. *Glencore total revenue 2020.* [online] Available at: <https://www.statista.com/statistics/274687/total-revenue-of-glencore-xstrata/> [Accessed 13 April 2021].

26. The Economist. 2013. *Marc Rich.* [online] Available at: <https://www.economist.com/obituary/2013/07/06/marc-rich> [Accessed 13 April 2021].

27. Benchmark Mineral Intelligence. 2019. *Glencore closes Mutanda mine, 20% of global cobalt supply comes offline.* [online] Available at: <https://www.benchmarkminerals.com/glencore-closes-mutanda-mine-20-of-global-cobalt-supply-comes-offline/> [Accessed 13 April 2021].

28. Grant, A., Hersh, E., Galli, C., Jimenez, D. and Brooker, M., 2020. *Is Lithium Brine Water?—Anti-Webinar Summary & Conclusions.* [online] Jade Cove Partners. Available at: <https://www.jadecove.com/research/islithiumbrinewaterantiwebinar> [Accessed 13 April 2021].

29. Sherwood, D., 2018. *Chilean regulators reject Albemarle's plans to boost lithium output.* [online] Reuters.com. Available at: <https://www.reuters.com/article/uk-chile-lithium-albemarle-exclusive-idUK-KCN1NI1Fd> [Accessed 13 April 2021].

30. Sgs.com. 2010. *HARD ROCK LITHIUM PROCESSING.* [online] Available at: <https://www.sgs.com/-/media/Global/Documents/Flyers%20and%20Leaflets/SGS-MIN-WA109-Hard-Rock-Lithium-Processing-EN-11.pdf> [Accessed 19 March 2021].

31. Dunn, J., Gaines, L., Barnes, M., Wang, M. and Sullivan, J., 2012. 'Material and energy flows in the materials production, assembly, and end-of-life stages of the automotive lithium-ion battery life cycle'.

32. Grant, A., Deak, D. and Pell, R., 2020. *The CO2 Impact of the 2020s Battery Quality Lithium Hydroxide Supply Chain.* [online] Jade Cove Partners. Available at: <https://www.jadecove.com/research/liohco-2impact> [Accessed 13 April 2021].

33. López, R., 2015. *Bolivia's lithium boom: dream or nightmare?*. [online] openDemocracy. Available at: <https://www.opendemocracy.net/en/democraciaabierta/bolivia-s-lithium-boom-dream-or-nightmare/> [Accessed 13 April 2021].

34. Young, E., 2019. *Enormous lithium waste dump plan shows how shamefully backward we are.* [online] The Sydney Morning Herald. Available at: <https://www.smh.com.au/national/enormous-lithium-waste-dump-plan-shows-how-shamefully-backward-we-are-20190621-p52054.html> [Accessed 13 April 2021].

35. Dunn, J., Gaines, L., Barnes, M., Wang, M. and Sullivan, J., 2012. 'Material and energy flows in the materials production, assembly, and end-of-life stages of the automotive lithium-ion battery life cycle'.

36. News.metal.com. 2020. *Output data of Battery Materials in China in 2019: lithium hydroxide is expected to break out in the contradiction between supply and demand of Lithium Salt_SMM | Shanghai Non ferrous Metals.* [online] Available at: <https://news.metal.com/news content/101017564/%5Bsmm-analysis%5D-output-data-of-battery-materials-in-china-in-2019:-lithium-hydroxide-is-expected-to-break-out-in-the-contradiction-between-supply-and-demand-of-lithium-salt/> [Accessed 13 April 2021].

37. McCarthy, N., 2018. *China Produces More Cement Than The Rest Of The World Combined [Infographic].* [online] Forbes. Available at: <https://www.forbes.com/sites/niallmccarthy/2018/07/06/china-produces-more-cement-than-the-rest-of-the-world-combined-infographic/?sh=74122ff36881> [Accessed 13 April 2021].

38. IEA. 2020. *Electricity mix in China, Q1 2020—Charts—Data & Statistics.* [online] Available at: <https://www.iea.org/data-and-statistics/charts/electricity-mix-in-china-q1–2020> [Accessed 13 April 2021].

39. Appunn, K., Haas, Y. and Wettengel, J., 2021. *Germany's energy consumption and power mix in charts.* [online] Clean Energy Wire. Available at: <https://www.cleanenergywire.org/factsheets/germanys-energy-consumption-and-power-mix-charts> [Accessed 13 April 2021].

40. Hickman, L., 2012. *Are electric cars bad for the environment? | Leo Hickman.* [online] The Guardian. Available at: <https://www.the-

guardian.com/environment/blog/2012/oct/05/electric-cars-emissions-bad-environment> [Accessed 13 April 2021].

41. Reichmuth, D., 2020. *Are Electric Vehicles Really Better for the Climate? Yes. Here's Why.* [online] Union of Concerned Scientists. Available at: <https://blog.ucsusa.org/dave-reichmuth/are-electric-vehicles-really-better-for-the-climate-yes-heres-why> [Accessed 19 March 2021].

42. Gerretsen, I., 2020. *Japan net zero emissions pledge puts coal in the spotlight.* [online] Climate Home News. Available at: <https://www.climatechangenews.com/2020/10/26/japan-net-zero-emissions-pledge-puts-coal-spotlight/#:~:text=Japan%20is%20the%20world's%20fifth,third%20of%20its%20electricity%20generation.> [Accessed 13 April 2021].

43. Tabuchi, H., 2020. *Japan Races to Build New Coal-Burning Power Plants, Despite the Climate Risks (Published 2020).* [online] Nytimes.com. Available at: <https://www.nytimes.com/2020/02/03/climate/japan-coal-fukushima.html> [Accessed 13 April 2021].

44. InsideEVs. 2020. *According To Tesla CEO Elon Musk, This Metal Is The New Gold.* [online] Available at: <https://insideevs.com/news/440582/elon-musk-lithium-ion-battery-nickel-is-new-gold/> [Accessed 13 April 2021].

45. Desjardins, J., 2017. *Nickel: The Secret Driver of the Battery Revolution.* [online] Visual Capitalist. Available at: <https://www.visualcapitalist.com/nickel-secret-driver-battery-revolution/> [Accessed 19 March 2021].

46. Verne, S. and Williams, J., 2021. *Natural Graphite Active Anode Material (AAM) for Global Electric Vehicle Demand.* [online] Syrahresources.com.au. Available at: <http://www.syrahresources.com.au/application/third_party/ckfinder/userfiles/files/20210120%20Advanced%20Automotive%20Battery%20Conference%20Presentation(1).pdf> [Accessed 13 April 2021].

47. Whoriskey, P., Robinson Chavez, M. and Ribas, J., 2016. *IN YOUR PHONE, IN THEIR AIR. A trace of graphite is in consumer tech. In these Chinese villages, it's everywhere.* [online] Available at: <https://www.washingtonpost.com/graphics/business/batteries/graphite-mining-pollution-in-china/> [Accessed 18 March 2021].

48. Ibid.
49. Norris, F., 2014. *In China, Detecting Fraud Riskier Than Doing It (Published 2014)*. [online] Nytimes.com. Available at: <https://www.nytimes.com/2014/08/29/business/in-china-detecting-fraud-riskier-than-doing-it.html> [Accessed 14 April 2021].
50. Luhn, A., 2016. *Where the river runs red: can Norilsk, Russia's most polluted city, come clean?*. [online] The Guardian. Available at: <https://www.theguardian.com/cities/2016/sep/15/norilsk-red-river-russias-most-polluted-city-clean> [Accessed 19 March 2021].
51. Ibid.
52. Reuters.com. 2019. *Indonesia hopes for environmental nod soon for battery-grade nickel plants*. [online] Available at: <https://www.reuters.com/article/us-indonesia-nickel-environment-idUSKBN1XS1Sq> [Accessed 14 April 2021].
53. Cruz, E., 2017. *Philippines' Duterte warns miners: 'I will tax you to death'*. [online] Reuters.com. Available at: <https://www.reuters.com/article/us-philippines-duterte-mining-idUSKBN1A90Xd> [Accessed 14 April 2021].
54. Almendral, A., 2017. *Philippines Moves to Shut Mines Accused of Polluting (Published 2017)*. [online] Nytimes.com. Available at: <https://www.nytimes.com/2017/04/27/world/asia/philippines-mining-environment.html> [Accessed 18 March 2021].
55. Morse, I., 2020. *Indonesian miners eyeing EV nickel boom seek to dump waste into the sea*. [online] Mongabay Environmental News. Available at: <https://news.mongabay.com/2020/05/indonesian-miners-eyeing-ev-nickel-boom-seek-to-dump-waste-into-the-sea/> [Accessed 18 March 2021].
56. Morse, I., 2019. *In Indonesia, a tourism village holds off a nickel mine— for now*. [online] Mongabay Environmental News. Available at: <https://news.mongabay.com/2019/12/in-indonesia-a-tourism-village-holds-off-a-nickel-mine-for-now/> [Accessed 18 March 2021].

6. URBAN MINING

1. Jacobs, J., 1970. *The economy of cities*. New York: Vintage Books, pp. 110–112.

2. Zhang, H., 2020. *Challenges to Making Lithium-ion Batteries and Electric Vehicles Environmentally Friendly.* [online] Center for Integrated Catalysis. Available at: <https://cicchemistry.com/2020/12/01/challenges-to-making-lithium-ion-batteries-and-electric-vehicles-environmentally-friendly/> [Accessed 14 April 2021].

3. Eckart, J., 2017. *Batteries can be part of the fight against climate change—if we do these five things.* [online] World Economic Forum. Available at: <https://www.weforum.org/agenda/2017/11/battery-batteries-electric-cars-carbon-sustainable-power-energy/> [Accessed 14 April 2021].

4. Yumae, S., 2020. *Resource-poor Japan unearths metal riches in its trash.* [online] Nikkei Asia. Available at: <https://asia.nikkei.com/Business/Markets/Commodities/Resource-poor-Japan-unearths-metal-riches-in-its-trash> [Accessed 19 March 2021].

5. Szymkowski, S., 2019. *Toyota will use Tokyo Olympics to debut solid-state battery electric vehicle.* [online] Roadshow. Available at: <https://www.cnet.com/roadshow/news/toyota-solid-state-battery-electric-olympics/> [Accessed 14 April 2021].

6. Stoklosa, A., 2019. *Toyota Has a Curious Justification for Not Selling Any EVs (Yet).* [online] Car and Driver. Available at: <https://www.caranddriver.com/news/a26703778/toyota-why-not-selling-electric-cars/> [Accessed 14 April 2021].

7. JX Nippon Mining & Metals. 2020. *Corporate History | Corporate Overview.* [online] Available at: <https://www.nmm.jx-group.co.jp/english/company/history.html> [Accessed 14 April 2021].

8. Haga, Y., Saito, K. and Hatano, K., 2018. Waste Lithium-Ion Battery Recycling in JX Nippon Mining & Metals Corporation. *The Minerals, Metals & Materials Series*, pp. 143–147.

9. JX Nippon Mining & Metals. 2020. *Corporate History | Corporate Overview.* [online] Available at: <https://www.nmm.jx-group.co.jp/english/company/history.html> [Accessed 14 April 2021].

10. Statista. n.d. *China: new energy vehicle sales by type 2020.* [online] Available at: <https://www.statista.com/statistics/425466/china-annual-new-energy-vehicle-sales-by-type/> [Accessed 14 April 2021].

11. Geman, B., 2020. *Global electric vehicle sales topped 2 million in 2019.* [online] Axios. Available at: <https://www.axios.com/electric-vehi-

cles-worldwide-sales-2fea9c70–411f-47d3–9ec6–487c7075482c.html>
[Accessed 14 April 2021].

12. Gu, T., 2019. *Newzoo's Global Mobile Market Report: Insights into the World's 3.2 Billion Smartphone Users, the Devices They Use & the Mobile Games They Play.* [online] Newzoo. Available at: <https://newzoo.com/insights/articles/newzoos-global-mobile-market-report-insights-into-the-worlds-3-2-billion-smartphone-users-the-devices-they-use-the-mobile-games-they-play/> [Accessed 14 April 2021].

13. Chinavitae.com. n.d. *China Vitae: Biography of Xiao Yaqing.* [online] Available at: <https://www.chinavitae.com/biography/Xiao_Yaqing> [Accessed 19 March 2021].

14. Batteryuniversity.com. n.d. *Types of Battery Cells; Cylindrical Cell, Button Cell, Pouch Cell.* [online] Available at: <https://batteryuniversity.com/learn/article/types_of_battery_cells> [Accessed 19 March 2021].

15. Oberhaus, D., 2020. *Where Was the Battery at Tesla's Battery Day?.* [online] Wired. Available at: <https://www.wired.com/story/where-was-the-battery-at-teslas-battery-day/> [Accessed 14 April 2021].

16. Fleming, W., 2019. *Cats vs Dogs—Part 1—92.8% Accuracy—Binary Image Classification with Keras and Deep Learning.* [online] Will Fleming's Software blog. Available at: <https://wtfleming.github.io/2019/05/07/keras-cats-vs-dogs-part-1/> [Accessed 19 March 2021].

17. Mint. 2019. *Meet Daisy, the new Apple robot who can disassemble 200 iPhones per hour.* [online] Available at: <https://www.livemint.com/technology/tech-news/meet-daisy-the-new-apple-robot-who-can-disassemble-200-iphones-per-hour-1555654439509.html> [Accessed 19 March 2021].

18. Wiens, K., 2016. *Apple's Recycling Robot Needs Your Help to Save the World.* [online] Wired. Available at: <https://www.wired.com/2016/03/apple-liam-robot/> [Accessed 19 March 2021].

19. Mint. 2019. *Meet Daisy, the new Apple robot who can disassemble 200 iPhones per hour.* [online] Available at: <https://www.livemint.com/technology/tech-news/meet-daisy-the-new-apple-robot-who-can-disassemble-200-iphones-per-hour-1555654439509.html> [Accessed 19 March 2021].

20. Trend Tracker. 2020. *IMI CEO and President issue an open letter on the electric vehicle repair skills gap.* [online] Available at: <https://www.trendtracker.co.uk/imi-ceo-and-president-issue-an-open-letter-on-the-electric-vehicle-repair-skills-gap/> [Accessed 19 March 2021].

21. En.gem.com.cn. n.d. *GEM Co., Ltd.* [online] Available at: <http://en.gem.com.cn/en/AboutTheGroup/index.html> [Accessed 14 April 2021].

22. Ibid.

23. Ellenmacarthurfoundation.org. n.d. 节约材料开启移动变革新篇章. [online] Available at: <https://www.ellenmacarthurfoundation.org/cn/%E6%A1%88%E4%BE%8B%E5%88%86%E6%9E%90/%E8%8A%82%E7%BA%A6%E6%9D%90%E6%96%99%E5%BC%80%E5%90%AF%E7%A7%BB%E5%8A%A8%E5%8F%98%E9%9D%A9%E6%96%B0%E7%AF%87%E7%AB%A0> [Accessed 14 April 2021].

24. Jacoby, M., 2019. *It's time to get serious about recycling lithium-ion batteries.* [online] Cen.acs.org. Available at: <https://cen.acs.org/materials/energy-storage/time-serious-recycling-lithium/97/i28> [Accessed 19 March 2021].

25. Csm.umicore.com. n.d. *Our recycling process.* [online] Available at: <https://csm.umicore.com/en/battery-recycling/our-recycling-process/> [Accessed 15 April 2021].

26. Offshore Energy. 2020. *Misdeclared Lithium Battery Cargo Caused Cosco Pacific's Fire—Offshore Energy.* [online] Available at: <https://www.offshore-energy.biz/misdeclared-lithium-battery-cargo-caused-cosco-pacifics-fire/> [Accessed 19 March 2021].

27. Warwick.ac.uk. 2020. *Automotive Lithium ion Battery Recycling in the UK.* [online] Available at: <https://warwick.ac.uk/fac/sci/wmg/business/transportelec/22350m_wmg_battery_recycling_report_v7.pdf> [Accessed 15 April 2021].

28. Roskill. 2019. *Batteries: GEM signs agreement to supply 170kt of raw materials to ECOPRO.* [online] Available at: <https://roskill.com/news/batteries-gem-signs-agreement-to-supply-170kt-of-raw-materials-to-ecopro/> [Accessed 19 March 2021].

29. Reuters.com. 2021. *China's GEM seeks to double stake, take control of*

Indonesia nickel project. [online] Available at: <https://www.reuters.com/article/gem-indonesia-nickel-cobalt-idUSL4N2JF0Ug> [Accessed 15 April 2021].

30. Radford, C., 2020. *Glencore, GEM extend cobalt supply deal until 2029.* [online] Metalbulletin.com. Available at: <https://www.metalbulletin.com/Article/3964964/Glencore-GEM-extend-cobalt-supply-deal-until-2029.html> [Accessed 15 April 2021].

31. South China Morning Post. 2019. *Meet the former professor behind GEM, the world's biggest battery recycler.* [online] Available at: <https://www.scmp.com/tech/big-tech/article/3039452/chinas-gem-worlds-biggest-battery-recycler-helping-fuel-future-cars> [Accessed 19 March 2021].

32. En.gem.com.cn. n.d. *GEM Co., Ltd.* [online] Available at: <http://en.gem.com.cn/en/UsedBatteryRecycling/index.html> [Accessed 15 April 2021].

33. Farchy, J. and Warren, H., 2018. *China Has a Secret Weapon in the Race to Dominate Electric Cars.* [online] Bloomberg.com. Available at: <https://www.bloomberg.com/graphics/2018-china-cobalt/?sref=TtblOutp> [Accessed 15 April 2021].

34. Ibid.

35. Deign, J., 2019. *How China Is Cornering the Lithium-Ion Cell Recycling Market.* [online] Greentechmedia.com. Available at: <https://www.greentechmedia.com/articles/read/how-china-is-cornering-the-lithium-ion-cell-recycling-market> [Accessed 19 March 2021].

36. Zhang, J., 2020. *China on track to hit target of building 500,000 5G base stations this year.* [online] South China Morning Post. Available at: <https://www.scmp.com/tech/gear/article/3100491/china-has-reached-about-96-cent-its-target-build-500000–5g-base-stations> [Accessed 15 April 2021].

7. BRIGHT GREEN FUTURE

1. Atag.org. n.d. *Facts & figures.* [online] Available at: <https://www.atag.org/facts-figures.html#:-:text=The%20global%20aviation%20industry%20produces,carbon%20dioxide%20(CO2)%20emissions.&text=Aviation%20is%20responsible%20for%2012,to%2074%25%20from%20road%20transport.> [Accessed 15 April 2021].

2. Transportenvironment.org. 2019. *Shipping and climate change.* [online] Available at: <https://www.transportenvironment.org/what-we-do/shipping-and-environment/shipping-and-climate-change#:~:text=The%20Third%20IMO%20GHG%20Study,of%20annual%20global%20CO2%20emissions.&text=Shipping%20also%20contributes%20to%20climate,by%20combustion%20of%20marine%20fuel.> [Accessed 22 March 2021].

3. Thomson, R., 2020. *Electric propulsion is finally on the map.* [online] Roland Berger. Available at: <https://www.rolandberger.com/en/Insights/Publications/Electric-propulsion-is-finally-on-the-map.html> [Accessed 15 April 2021].

4. Kane, M., 2020. *China: In Some BEVs, Battery Cell Energy Density Now Reaches 250–280 Wh/kg.* [online] InsideEVs. Available at: <https://insideevs.com/news/428511/china-battery-energy-density-280-wh-kg/> [Accessed 15 April 2021].

5. Burke, A. and Miller, M., 2010. *The UC Davis Emerging Lithium Battery Test Project.* [online] Escholarship.org. Available at: <https://escholarship.org/uc/item/4xn6n3xf> [Accessed 15 April 2021].

6. Anderson, D. and Patiño-Echeverri, D., 2009. 'An Evaluation of Current and Future Costs for Lithium-Ion Batteries for Use in Electrified Vehicle Powertrains'. *Nicholas School of the Environment of Duke University.*

7. Writer, B., 2019. *Lithium-Ion Batteries A Machine-Generated Summary of Current Research.* Cham, Switzerland: Springer, pp. 1–10.

8. Airbus. 2019. *Thermal engines vs. electric motors.* [online] Available at: <https://www.airbus.com/newsroom/stories/airbus-pursues-hybrid-propulsion-solutions-for-future-air-vehicles.html> [Accessed 15 April 2021].

9. Aeromontreal.ca. n.d. *Omer Bar-Yohay.* [online] Available at: <https://www.aeromontreal.ca/fiche-omer-bar-yohay-574.html> [Accessed 15 April 2021].

10. Eviation.co. n.d. *Aircraft—Eviation.* [online] Available at: <https://www.eviation.co/aircraft/#Alice-Specifications> [Accessed 15 April 2021].

11. Cantu, M., 2021. *Charging Speed Race: Tesla Model 3 Vs Audi E-Tron.*

[online] InsideEVs. Available at: <https://insideevs.com/news/494787/charging-speed-race-model-3-audi-etron/> [Accessed 15 April 2021].

12. Randall, C., 2018. *Eviation Aircraft sets sights on Kokam batteries.* [online] Electrive.com. Available at: <https://www.electrive.com/2018/02/15/eviation-aircraft-sets-sights-kokam-batteries/> [Accessed 23 March 2021].

13. Ibid.

14. Jasper, C., 2019. *Eviation Lands More Customers as Electric Plane Orders Top 150.* [online] Bloomberg.com. Available at: <https://www.bloomberg.com/news/articles/2019-10-24/eviation-lands-more-customers-as-electric-plane-orders-top-150?sref=TtblOutp> [Accessed 15 April 2021].

15. Bellamy III, W., 2019. *Eviation CEO: Alice to Start Electric Powered Flights by 2021.* [online] Aviation Today. Available at: <https://www.aviationtoday.com/2019/01/25/eviation-ceo-alice-start-electric-powered-flights-2021/> [Accessed 23 March 2021].

16. Alcock, C., 2020. *Eviation's Electric Alice Aircraft Catches Fire During Ground Tests.* [online] Aviation International News. Available at: <https://www.ainonline.com/aviation-news/business-aviation/2020-01-24/eviations-electric-alice-aircraft-catches-fire-during-ground-tests#:~:text=A%20fire%20broke%20out%20during,a%20ground%2Dbased%20battery%20system.&text=Eviation%20pushed%20back%20plans%20to,in%20late%202019%20into%202020.> [Accessed 23 March 2021].

17. Airbus. 2017. *Airbus, Rolls-Royce, and Siemens team up for electric future Partnership launches E-Fan X hybrid-electric flight demonstrator.* [online] Available at: <https://www.airbus.com/newsroom/press-releases/en/2017/11/airbus--rolls-royce--and-siemens-team-up-for-electric-future-par.html> [Accessed 15 April 2021].

18. Rochesteravionicarchives.co.uk. n.d. *BAe 146.* [online] Available at: <https://rochesteravionicarchives.co.uk/platforms/bae-146#:~:text=With%20387%20aircraft%20produced%2C%20the,has%20retractable%20tricycle%20landing%20gear.> [Accessed 15 April 2021].

19. Aerospace-technology.com. n.d. *E-Fan X Hybrid-Electric Propulsion Aircraft, France.* [online] Available at: <https://www.aerospace-tech-

nology.com/projects/e-fan-x-hybrid-electric-aircraft/>   [Accessed 15 April 2021].

20. Excell, J., 2020. *Rolls-Royce and Airbus cancel E-Fan X project.* [online] The Engineer. Available at: <https://www.theengineer.co.uk/e-fan-x-project-cancelled/#:~:text=E%2DFan%20X%2C%20a%20joint,it's%20maiden%20flight%20in%202021.> [Accessed 23 March 2021].

21. BBC News. 2019. *'World's first' fully-electric commercial flight takes off.* [online] Available at: <https://www.bbc.com/news/business-50738983> [Accessed 15 April 2021].

22. Morris, C., 2020. *Electric Cessna Grand Caravan makes historic maiden flight.* [online] Charged EVs. Available at: <https://chargedevs.com/newswire/electric-cessna-grand-caravan-makes-historic-maiden-flight/> [Accessed 23 March 2021].

23. Quanlin, Q., 2017. *Fully electric cargo ship launched in Guangzhou—Business.* [online] Chinadaily.com.cn. Available at: <https://www.chinadaily.com.cn/business/2017-11/14/content_34511312.htm> [Accessed 22 March 2021].

24. McFadden, C., 2020. *The Countries With the Most Electric Vehicle Owners.* [online] Interestingengineering.com. Available at: <https://interestingengineering.com/the-countries-with-the-most-electric-vehicle-owners#:~:text=Which%20country%20has%20the%20most%20EVs%20per%20population%3F,according%20to%20figures%20from%202018.> [Accessed 15 April 2021].

25. Kongsberg.com. 2017. *YARA and KONGSBERG enter into partnership to build world's first autonomous and zero emissions ship.* [online] Available at: <https://www.kongsberg.com/maritime/about-us/news-and-media/news-archive/2017/yara-and-kongsberg-enter-into-partnership-to-build-worlds-first-autonomous-and/> [Accessed 15 April 2021].

26. The Maritime Executive. 2020. *Construction of Yara Birkeland Paused.* [online] Available at: <https://www.maritime-executive.com/article/construction-of-yara-birkeland-paused#:~:text=Due%20to%20the%20Covid%2D19,steps%20together%20with%20its%20partners.> [Accessed 15 April 2021].

27. Rodrigue, J., n.d. *Fuel Consumption by Containership Size and Speed* |

The Geography of Transport Systems. [online] Transportgeography.org. Available at: <https://transportgeography.org/contents/chapter4/transportation-and-energy/fuel-consumption-containerships/#:-:text=Fuel%20consumption%20by%20a%20containership,per%20day%20at%2024%20knots.> [Accessed 15 April 2021].

28. Boloor, M., Valderrama, P., Statler, A. and Garcia, S., 2019. *Electric Vehicles 101*. [online] NRDC. Available at: <https://www.nrdc.org/experts/madhur-boloor/electric-vehicles-101> [Accessed 15 April 2021].

29. Spbes.com. n.d. *Lithium NMC Marine Batteries*. [online] Available at: <https://spbes.com/products/> [Accessed 15 April 2021].

30. MIT Club of Northern California, 2019. *The Future of Energy Storage—Professor Yet-Ming Chiang, MIT*. [video] Available at: <https://www.youtube.com/watch?v=E76q-9q7ZDg> [Accessed 22 March 2021].

31. Google Tech Talks, 2018. *Post and Beyond Lithium-Ion Materials and Cells for Electrochemical Energy Storage*. [video] Available at: <https://www.youtube.com/watch?v=pxC2pciLl04&t=2304s> [Accessed 22 March 2021].

32. Crider, J., 2020. *Tesla Air? Elon Musk Hints Tesla Could Mass Produce 400 Wh/kg Batteries In 3–4 Years*. [online] CleanTechnica. Available at: <https://cleantechnica.com/2020/08/25/tesla-air-elon-musk-hints-tesla-could-mass-produce-400-wh-kg-batteries-in-3-4-years/> [Accessed 15 April 2021].

33. Energystorage.pnnl.gov. n.d. *PNNL: Energy Storage: Battery500*. [online] Available at: <https://energystorage.pnnl.gov/battery500.asp> [Accessed 15 April 2021].

34. Hall, M., 2020. *Energy density advances and faster charging would unlock EV revolution*. [online] pv magazine. Available at: <https://www.pv-magazine.com/2020/02/11/energy-density-advances-and-faster-charging-would-unlock-ev-revolution/> [Accessed 15 April 2021].

35. BloombergNEF. 2020. *Battery Pack Prices Cited Below $100/kWh for the First Time in 2020, While Market Average Sits at $137/kWh*. [online] Available at: <https://about.bnef.com/blog/battery-pack-prices-cited-below-100-kwh-for-the-first-time-in-2020-while-market-average-sits-at-137-kwh/> [Accessed 15 April 2021].

36. Keen, K., 2020. *As battery costs plummet, lithium-ion innovation hits lim-*

its, experts say. [online] Spglobal.com. Available at: <https://www.spglobal.com/marketintelligence/en/news-insights/latest-news-headlines/as-battery-costs-plummet-lithium-ion-innovation-hits-limits-experts-say-58613238> [Accessed 15 April 2021].

37. The Limiting Factor, 2020. *Professor Shirley Meng: The Future of the Anode (C, Si, Li).* [video] Available at: <https://www.youtube.com/watch?v=0ktsgwzUh3a> [Accessed 15 April 2021].

38. Ibid.

39. Evarts, E., 2015. 'Lithium batteries: To the limits of lithium'. *Nature,* 526(7575), pp. S93–S95.

INDEX

INDEX

INDEX

INDEX

INDEX

INDEX

INDEX

INDEX

INDEX

INDEX